ELEMENTS OF LATTICE THEORY

L. A. Skornjakov

ELEMENTS OF LATTICE THEORY

Translated and Edited from the Russian by V. Kumar

HINDUSTAN PUBLISHING CORPORATION (INDIA) DELHI
1977
ADAM HILGER LTD BRISTOL

ELEMENTS OF LATTICE THEORY
Title of Russian Original Edition : *Elementi Teorii Struktur*

Distributed in the Western Hemisphere, Europe, U. K., South Africa, Australia by Adam Hilger Ltd., Bristol, United Kingdom

British Library Cataloguing in Publication Data
Skornjakov, Lev Anatol'evich
Elements of lattice theory
1. Lattice Theory
I. Title II. Kumar, V
512'.7 QA171.5
ISBN 0 85274 331 9

Printed in India by
Hindustan Publishing Corporation Printing Press (India), Delhi-110007

PREFACE

The present text is addressed to those mathematicians who want to become acquainted with the basics of lattice theory, though they may have not yet (or already!) chosen this branch of mathematics for their specialization. Hence our choice is biased in favour of those results which either stimulate further development of lattice-theoretic ideas or find applications in other fields of mathematics. For a deeper and more intimate knowledge of lattice theory, the works of Birkhoff and Sikorski should be referred to. The subsequent achievements in lattice theory are illuminated in the relevant papers in [6]. The requisite bibliography is appended at the end of the book. The textbooks on lattice theory known to us are also listed there. Some of these texts are accompanied with a list of excellent and detailed references from the periodical literature. However, we have restricted the listing here to those references which have a bearing on some of the conclusions included in the text.

No prior knowledge of the subject is demanded from the reader. A grasp of the fundamentals of elementary set theory would suffice. Though in the text we frequently speak of a group, ring and universal algebra, these expressions bear the character of a 'literary digression'. The set-theoretic symbols of union $\cup$ and intersection $\cap$ are employed in the usual sense. The set B when it is a complement of the set A is denoted by $A \setminus B$. The symbols $\subseteq$ and $\in$ are used in the customary sense.

By the direct product $A \times B$ we understand the set of all pairs (a, b), where $a \in A$ and $b \in B$. This notion is naturally extended to an arbitrary number of terms. The logical difficulties arising in this connection are discussed in Chapter 2.

We write

$$\{x \mid x \text{ having the property } \mathcal{E}\}$$

to denote a set of all elements x having the property $\mathcal{E}$. The symbol

$$\underset{\text{def}}{=}$$

stands for the words 'is equal to by definition'. When reference is made to a theorem from a different chapter, a chapter numeral is prefixed to the number, e.g. Theorem 5.3. refers to Theorem 3 from Chapter 5. The presence of only one number indicates that the requisite theorem is within the same chapter.

The manuscript was read by A. H. Livshitz, who made several valuable suggestions. Considerable improvement in the presentation is due also to critical comments of T. S. Fofanov who, in addition, rechecked all the exercises. The author is deeply indebted to these mathematicians.

CONTENTS

PREFACE v

1. PARTIALLY ORDERED SETS 1

2. ORDINAL NUMBERS 12

3. COMPLETE LATTICES 29

4. LATTICES 41

5. FREE LATTICES 58

6. MODULAR LATTICES 90

7. DISTRIBUTIVE LATTICES 117

8. BOOLEAN ALGEBRAS 129

BIBLIOGRAPHY 143

INDEX 145

LIST OF SPECIAL SYMBOLS 148

1

PARTIALLY ORDERED SETS

Let P be an arbitrary non-void set. A *relation* on the set P is defined as a subset ρ of the direct product $P \times P$. If $(a, b) \in \rho$, we say that the elements a and b stand in the relation ρ, and we write $a\rho b$. The relation consisting of all pairs (a, a), where $a \in P$, is called the *diagonal* and denoted by Δ. The relation coinciding with the entire set $P \times P$ is called the *identity*.

If P' is a subset of the set P, on which the relation ρ is defined, then we say that the relation ρ' on P' given by the condition

$$a\rho' b \underset{\text{def}}{=} a\rho b$$

is *induced by the relation* ρ.

A relation ρ is said to be :

(i) *reflexive*, if $a\rho a$ for every $a \in P$. In other words ρ is reflexive, if $\Delta \subseteq \rho$;

(ii) *symmetric*, if $a\rho b$ implies $b\rho a$;

(iii) *antisymmetric*, if $a\rho b$ and $b\rho a$ imply $a = b$;

(iv) *transitive*, if $a\rho b$ and $b\rho c$ imply $a\rho c$.

It is easy to verify that the diagonal has all the properties listed above. The identity relation has all of them but the antisymmetric property.

However, if P consists only of a single element, then the identity relation coincides with the diagonal.

A relation which is reflexive, symmetric and transitive is called an *equivalence relation* or simply an *equivalence*. The notion of equivalence plays an important role in almost all branches of mathematics, since it is related to the partitioning of a set P, that is, to its representation in the form of a union of mutually disjoint subsets. To clarify this connection, the set of all x in P such that $x\rho a$ is said to be the *residue class* or *block* of ρ defined by the element $a \in P$. It is clear that P coincides with the union of all blocks of every reflexive relation. A system of subsets of P is called a *partition* if these subsets are mutually disjoint and their union coincides with P.

THEOREM 1. *The collection of blocks of an equivalence ρ on a set P is a partition.*

Proof. In view of the reflexivity, every element of P is contained in the block defined by it. Now, consider the blocks A and B, defined by the elements a and b, respectively. If $x \in A$, $y \in B$ and $z \in A \cap B$, then we have $x\rho a$, $z\rho a$, $z\rho b$ and $y\rho b$. Since the relation ρ is symmetric, we have the chains

$$x\rho a,\ a\rho z,\ z\rho b$$

and

$$y\rho b,\ b\rho z,\ z\rho a,$$

whence, by transitivity, it follows that $x\rho b$ and $y\rho a$, that is, $A \subseteq B \subseteq A$. Thus, distinct blocks are disjoint.

A set of blocks of an equivalence ρ, defined on a set P, is called a *factor set* of P with respect to ρ and is denoted by P/ρ. The mapping P on P/ρ, establishing a correspondence between each element of P and the block defined by it is called *natural*.

Conversely, every partition of a set P defines an equivalence on P, for the next assertion is immediate from the definition.

THEOREM 2. *If Σ is a partition of P, then the relation*

$$a\rho b \underset{\text{def}}{=} \boxed{\begin{array}{l}a \text{ and } b \text{ belong to one and}\\ \text{the same subset of } \Sigma\end{array}}$$

is an equivalence, whose blocks are just the subsets of the system Σ.

Note that if, starting from an equivalence ρ, we construct by Theorem 1 a partition Σ and then consider the relation described in Theorem 2, we obtain again the relation ρ. In exactly the same way, if starting from some partition Σ, we successively effect the constructions described in Theorems 2 and 1, we again reach Σ. Thus, there is a one-to-one correspondence between the equivalences on the given set and its partitions.

Our next object of study is an *order* relation, that is, a relation which is reflexive, antisymmetric and transitive. Any order relation evidently contains the diagonal. The inclusion relation on the set $\mathfrak{P}(\mathfrak{M})$ of all subsets of a set $\mathfrak{M}$ is also an order relation.

THEOREM 3. *If $\leqslant$ is a reflexive and transitive relation on the set P, then the relation ρ :*

$$a\rho b \underset{\text{def}}{=} \boxed{a \leqslant b \text{ and } b \leqslant a}$$

is an equivalence. The relation $\leqslant$ on the factor set $\overline{P} = P/\rho$, defined by the condition

$$\overline{a} \leqslant \overline{b} \underset{\text{def}}{=} a \leqslant b$$

($\overline{a}$ is the block of ρ defined by a), *is an order relation.*

Proof. The reflexivity and transitivity of ρ are immediate from the corresponding properties of the relation $\leqslant$. The symmetry of ρ follows directly from the definition. If $\overline{a} = \overline{c}$, $\overline{b} = \overline{d}$ and $\overline{a} \leqslant \overline{b}$, then

$$c \leqslant a \leqslant b \leqslant d,$$

that is, $\overline{c} \leqslant \overline{d}$ and the relation $\leqslant$ on $\overline{P}$ is defined correctly. Its reflexivity and transitivity are easily deduced from the corresponding properties of $\leqslant$ on P. If $\overline{a} \leqslant \overline{b}$ and $\overline{b} \leqslant \overline{a}$, then

$$a \leqslant b \leqslant a,$$

that is, $\overline{a} = \overline{b}$. Thus, the relation $\leqslant$ is antisymmetric too.

A set P is called a *partially ordered set* (hereafter abridged to *poset*) if some order, usually denoted by the symbol $\leqslant$, is defined on it: the notation $a \leqslant b$ is read as 'a is less than or equal to b'. If $a \leqslant b$ and $a \neq b$, we write $a < b$ and say that a is less than b. It is clear that different order relations can be defined on one and the same set. The posets arising in this case are regarded as distinct. A set with the diagonal defined on it is called a *trivial* poset. An important poset is the set $\mathfrak{P}(\mathfrak{M})$, ordered by inclusion. *Every subset of a poset is, evidently, itself a poset relative to the induced order.*

Two posets P and P' are called *isomorphic* if there is a one-to-one mapping φ of P onto P', such that $a \leqslant b$ holds iff $\varphi(a) \leqslant \varphi(b)$. The fact that φ is a one-to-one mapping can be deduced from the latter condition. The mapping φ of a poset P into a poset P' is called *isotone* if $a \leqslant b$ implies $\varphi(a) \leqslant \varphi(b)$.

The elements a and b of a poset P are called *comparable*, if $a \leqslant b$ or $b \leqslant a$ holds. Two elements of a trivial poset are evidently comparable iff they coincide. An element v of a poset P is called the *greatest* element of P if $v \geqslant x$ for all $x \in P$. If $x \geqslant u$ for all $x \in P$, then u is called the *least* element of P. A greatest element is often called the *unit* element, with notation u, U or I, and a least element the *zero* element, with notation o, O or 0. A poset may of course contain neither a zero element nor a unit element. In particular, this is so for a trivial poset of not more than one element. An element M of a poset P is called a *maximal* element of P if $a \geqslant M$ for some $a \in P$ implies $a = M$. If $a \leqslant m$ for some $a \in P$ implies $a = m$, then m is called a *minimal* element of P. It is easy to verify that a greatest element is maximal and a least element minimal. The converse in general is not true. Thus, for example, in a trivial poset every element is both maximal and minimal. Another instance when the minimal elements are not the least elements is furnished by prime numbers in a set of positive integers, different from 1, with divisibility as order relation (that is, $m \leqslant n \underset{\text{def}}{=} m$ divides n).

Note that the definition of a least element is obtained from that of a greatest element simply by replacing the symbol $\leqslant$ by $\geqslant$. In exactly the same way, the notions of minimal and maximal elements are connected. In general, if in some expression of a poset we replace $\leqslant$ by $\geqslant$ we get a new expression, called the *dual*. Let P be a poset, and on P define the relation $\trianglelefteq$ by putting

$$a \trianglelefteq b \underset{\text{def}}{=} b \leqslant a.$$

It is easy to verify that $\trianglelefteq$ is an order relation. Thus, the set P becomes a poset P^*. Evidently, $P^{**} = P$. The posets P and P^* are said to be *mutually dual.*

THEOREM 3$'$. (Duality Principle). *If some theorem formulated in general logical and ordering terminology is true for all posets, then so is its dual.*

Proof. In fact, suppose that some theorem T is true. Assume that the conditions of duality of T^* are satisfied in some poset P. But, then in the dual poset P^*, the hypotheses of theorem $T^{**} = T$ are satisfied and, consequently, the conclusion of this theorem holds. Thus in the poset $P^{**} = P$ the conclusion of theorem T^* is true.

It may be emphasized again that for applying the duality principle, it is necessary to replace by duals all expressions and all notions appearing in the formulation and involving order relations, without affecting the general logical terms. In doubtful cases, one should either reproduce the proof of the duality principle or develop the argument, replacing every step in the proof of the theorem by its dual.

Let A be a non-void subset of a poset P, then the element $a \in P$ is called a *least upper bound* of A, if $a \geqslant x$ for all $x \in A$ and the relation $v \geqslant x$ for all $x \in A$ implies $v \geqslant a$. The greatest lower bound of A is defined dually as follows : an element $a \in P$ is called a *greatest lower bound* of A, if $a \leqslant x$ for all $x \in A$ and $u \leqslant x$ for all $x \in A$ implies $u \leqslant a$. The least upper and greatest lower bounds of A in a poset P are denoted by the symbols $\sup_P A$ and $\inf_P A$, respectively. However, the subscript P is usually dropped. The following assertions are now immediate :

(*a*) *If* $a \leqslant b$, *then* $\sup \{a, b\} = b$ *and* $\inf \{a, b\} = a$.
(*b*) *Let* $A \subseteq B \subseteq P$. *If there exist* $\sup A$ *and* $\sup B$ [$\inf A$ *and* $\inf B$], *then* $\sup A \leqslant \sup B$ [$\inf A \geqslant \inf B$].
(*c*) *If* $A \subseteq \mathfrak{P}(\mathfrak{M})$, *then* $\sup A$ *coincides with the union of the subsets, contained in* A, *and* $\inf A$ *with their intersection.*

These facts will be used here without special mention.

THEOREM 4. *If* $\{A_\alpha\}$ *is a set of subsets of a poset* P, $A = \bigcup A_\alpha$ *and if there exist* $\sup A$ *and* $\sup A_\alpha$ ($\inf A$ *and* $\inf A_\alpha$) *for all* α, *then*

$$\sup A = \sup \{\sup A_\alpha\} \ [\inf A = \inf \{\inf A_\alpha\}].$$

Proof. Put $a = \sup A$ and $a_\alpha = \sup A_\alpha$. Since $a \geqslant x_\alpha$ for all $x_\alpha \in A_\alpha$, it follows $a \geqslant a_\alpha$ for all α. If $v \geqslant a_\alpha$ for all α, then $v \geqslant x_\alpha$ for all $x_\alpha \in A_\alpha$ and, consequently, $v \geqslant x$ for all $x \in A$. Hence, $v \geqslant a$

and recalling the definition we find

$$\sup A = a = \sup a_\alpha = \sup \{\sup A_\alpha\}.$$

The second statement is proved by a dual argument.

THEOREM 5. *Let φ be an isotone mapping of a poset P into a poset P' and let A be a subset of P. If $a = \sup_P A$ and $a' = \sup_{P'} \varphi(A)$ exist, then $\varphi(a) \geqslant a'$. If $b = \inf_P A$ and $b' = \inf_{P'} \varphi(A)$ exist, then $\varphi(b) \leqslant b'$.*

Proof. Since $a \geqslant x$ for all $x \in A$, it follows that $\varphi(a) \geqslant \varphi(x)$ for all $x \in A$. Thus, $\varphi(a) \geqslant x'$ for all $x' \in \varphi(A)$, whence $\varphi(a) \geqslant a'$. The proof of the second assertion follows by duality.

Again consider a non-void subset of a poset A. The set of all $x \in P$ such that $x \geqslant a$ for all $a \in A$ is called the *upper cone* of A. We define the *lower cone* by duality. The upper and lower cones of A are denoted by A^Δ and A^∇, respectively.

THEOREM 6. *The upper and lower cones have the following properties:*

(i) *If $A \subseteq B$, then $A^\Delta \supseteq B^\Delta$ and $A^\nabla \supseteq B^\nabla$;*

(ii) $A \subseteq A^{\Delta\nabla} \cap A^{\nabla\Delta}$;

(iii) $A^\Delta = A^{\Delta\nabla\Delta}$;

(iv) $A^\nabla = A^{\nabla\Delta\nabla}$;

(v) $(A \cup B)^\Delta = A^\Delta \cap B^\Delta$;

(vi) $(A \cup B)^\nabla = A^\nabla \cap B^\nabla$.

Proof. Property (i) is immediate from definition. Since $x \leqslant y$ for every $x \in A$ and $y \in A^\Delta$, it follows that $A \subseteq A^{\Delta\nabla}$. Similarly, it is verified that $A \subseteq A^{\nabla\Delta}$. From (i) and (ii) and taking note of the associativity of the product of mappings, we obtain

$$A^\Delta \subseteq A^{\Delta(\nabla\Delta)} = A^{(\Delta\nabla)\Delta} \subseteq A^\Delta .$$

Property (iv) is proved similarly. The inclusion $(A \cup B)^\Delta \subseteq A^\Delta \cap B^\Delta$ follows from (i). Moreover, if $x \in A^\Delta \cap B^\Delta$, then $x \geqslant y$ for every $y \in A \cup B$, establishing the validity of property (v). Property (vi) is verified similarly.

THEOREM 7. *If* $\sup A$ *or* $\inf A^\Delta$ *exists, then*

$$\sup A = \inf A^{\Delta} .$$

Proof. If $a = \sup A$, then $x \leqslant a$ for all $x \in A$ and, in addition, $a \leqslant y$ for all $y \in A^{\Delta}$. However, if $u \leqslant y$ for all $y \in A^{\Delta}$, then, since $a \in A^{\Delta}$, we have $u \leqslant a$, verifying also the equality $a = \inf A^{\Delta}$. If there exists $b = \inf A^{\Delta}$, then, since for every $x \in A$ the relation $x \leqslant y$ holds for all $y \in A^{\Delta}$, we have $x \leqslant b$ for all $x \in A$. However, if $v \geqslant x$ for all $x \in A$, then $v \in A^{\Delta}$ and, consequently, $v \geqslant b$. Thus, $b = \sup A$.

Dually we have the next theorem.

THEOREM 8. *If* $\inf A$ *or* $\sup A^{\nabla}$ *exists, then*

$$\inf A = \sup A^{\nabla} .$$

If a and b are elements of a poset P, where $a \leqslant b$, then the set

$$[a, b] \underset{\text{def}}{=} \{x \mid a \leqslant x \leqslant b\}$$

is called the *interval* $[a, b]$. Of course, $a, b \in [a, b]$. An interval containing exactly two distinct elements is said to be *simple*.

Let Q be a subset of a poset P. As already remarked, Q is a poset. If $A \subseteq Q$ and there exists $a = \sup_P A \in Q$, then it is easy to verify that $a = \sup_Q A$. Similarly, from the existence of $a = \inf_P A \in Q$ it follows that $a = \inf_Q A$. Conversely, the existence of $\sup_Q A$, in general, does not imply the existence of $\sup_P A$ (Figure 1)†. What is more, the least

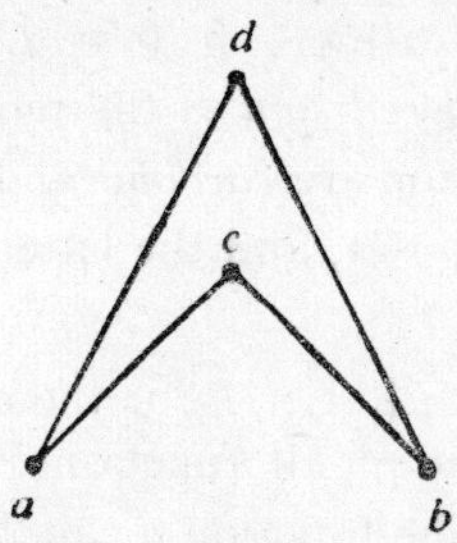

$P = \{a, b, c, d\}$,
$Q = \{a, b, c\}$,
$\sup_Q \{a, b\} = c$,
$\sup_P \{a, b\}$ does not exist

Figure 1.

upper bounds of $\sup_P A$ and $\sup_Q A$ may exist but do not coincide. This situation will be encountered in Chapter 3.

†In the graphical representation of a poset in Fig. 1, dark dots represent the elements of this set, which are joined in case they are the ends of a simple interval. Moreover, a larger element is placed higher than the smaller one. Thus, Figure 1 shows the case where $a < c$, $a < d$, $b < c$ and $b < d$.

A poset P is called a *chain* if any two elements in P are comparable. It is easy to convince oneself that every maximal (minimal) element of a chain is a greatest (least) element. It is trivial to verify that the join of an increasing nested sequence of chains of some poset is a chain. A chain $a_0 \leqslant a_1 \leqslant \ldots \leqslant a_n$ belonging to a poset with zero and unit elements is called a *composition series*, if $a_0 = 0$, $a_n = 1$ and all the intervals $[a_{i-1}, a_i]$ $(i = 1, \ldots, n)$ are simple. Obviously, a composition series does not always exist. The number n is called the *length* of the composition series.

Now, consider a system $\{P_\alpha \mid \alpha \in L\}$ of posets, assuming that the set L is also partially ordered. We agree to regard the distinct sets of the system as having no common point, notwithstanding the fact that some of them may appear to be distinct copies of the same set.

Let P be a set-theoretical union of the sets of the system $\{P_\alpha\}$. If $a, b \in P$, we put

$$a \leqslant b \underset{\text{def}}{=} \boxed{\begin{array}{c} a, b \in P_\alpha \text{ and } a \leqslant_{P_\alpha} b \\ \text{or} \\ a \in P_\alpha,\ b \in P_\beta \text{ and } \alpha < \beta \end{array}}\ \dagger$$

Clearly, $\leqslant$ is a reflexive relation. If $a \leqslant b$ and $b \leqslant a$ then, evidently, $a, b \in P_\alpha$ for some α and, consequently, $a = b$. If $a \leqslant b$ and $b \leqslant c$, then $a \in P_\alpha$, $b \in P_\beta$, $c \in P_\gamma$ and the inequality $a \leqslant c$ is established by a simple consideration of the following four cases : (*a*) $\alpha = \beta$, $\beta = \gamma$; (*b*) $\alpha = \beta$, $\beta < \gamma$; (*c*) $\alpha < \beta$, $\beta < \gamma$; (*d*) $\alpha < \beta$, $\beta = \gamma$. Thus, P is found to be a poset and is called an *ordered sum* of the sets P_α.

Two important cases of this construction are *cardinal* and *ordinal sums*. The former is obtained if L is a trivial poset, and the latter, if L forms a chain.

Consider again the system $\{P_\alpha \mid \alpha \in L\}$. Let $\overline{P}$ be the direct product of the sets of this system, that is, the set of all functions a, associating to every set P_α an element $a_\alpha \in P_\alpha$. The function a can be interpreted as the row (a_α), where α ranges over the set L. The problem of the existence of the set $\overline{P}$ when L is an infinite set is discussed in the next chapter. We define the relation $\leqslant$ on the set $\overline{P}$, (if it exists), by putting

$$(a_\alpha) \leqslant (b_\alpha) \underset{\text{def}}{=} \boxed{a_\alpha \leqslant b_\alpha \text{ for all } \alpha.}$$

†The index P_α under the symbol $\leqslant_{P_\alpha}$ indicates the poset, in which this relation is considered.

It is easy to verify that $\leqslant$ is an order relation. The poset obtained is called a *direct product* of the sets P_α. Though the order relation on L plays no role in the definition of a direct product, it is significant for other order relations defined on $\overline{P}$. We shall dwell upon this in the next chapter.

An ordered sum P contains summands in a natural way as subsets. This allows us to say that P decomposes into an ordered sum of its subsets. A poset not representable in the form of an ordinal (cardinal) sum of its subsets, is said to be *ordinally* (*cardinally*) *indecomposable*.

By contrast, the factors P_α of a direct product $\overline{P}$ do not admit such a natural interpretation as its subsets. It is true that if we consider a set of all rows in $\overline{P}$, having fixed elements at all places except some α_0, then we have a poset isomorphic to P_{α_0}. Thus, P_{α_0} is included in $\overline{P}$. However, such inclusions are many and each is equivalent to the other.

THEOREM 9. *Every poset is a cardinal sum of its cardinally indecomposable subsets.*

Proof. Consider on a given poset P the relation ρ, by putting

$$a\rho b \underset{\text{def}}{=} \boxed{\begin{array}{l} \text{There is a finite chain} \\ a = a_1, a_2, \ldots, a_{m-1}, a_m = b \\ \text{of elements in } P, \text{ such that } a_i \text{ and } a_{i+1} \\ \text{are comparable for } i = 1, 2, \ldots, m-1 \end{array}}$$

It is easy to see that ρ is an equivalence relation. By Theorem 1, $P = \bigcup P_\alpha$, where P_α are the blocks of ρ, and for $\alpha \neq \beta$ the intersection $P_\alpha \cap P_\beta$ is void and the elements $x \in P_\alpha$ and $y \in P_\beta$ are not comparable. Consequently, P is the cardinal sum of the subsets P_α. If for some α, a poset P_α is the cardinal sum of its non-void subsets A and B, then we choose $a \in A$ and $b \in B$. By definition P_α contains the sequence

$$a = a_1, a_2, \ldots, a_{m-1}, a_m = b,$$

where neighbouring elements are comparable. However, for some index i we have $a_i \in A$ and $a_{i+1} \in B$, which by the definition of a cardinal sum is not compatible with the elements a_i and a_{i+1} being comparable.

An analogous theorem for the ordinal sum also holds. This will be proved in Chapter 3.

Exercises

1. If the diagonal and the identity relation on a set P coincide, then P is a one-element set.
2. If a relation is symmetric and antisymmetric, then it is the diagonal.
3. Consider the following relations :

No.	*Set*	*Relation $A\rho B$*
1	All subsets of some set	A lies in B.
2	All finite subsets of some set	A and B contain the same number of elements
3	All subsets of some set	$A \cap B$ is non-void
4	All positive integers	The greatest common divisor of A and B is 1.
5	All positive integers	$A = B$ or $A = 2B$

Verify the entries in the following table [the sign + (or−) denotes that the corresponding property is satisfied (or not satisfied)].

No.	*Reflexivity*	*Symmetry*	*Antisymmetry*	*Transitivity*
1	+	−	+	+
2	+	+	−	+
3	+	+	−	−
4	−	+	−	−
5	+	−	+	−

4. A transitive relation ρ over P has the following properties : (*a*) $x\rho x$ is not satisfied for any $x \in P$; (*b*) if $x\rho y$, then $y \not\rho x$ (that is, $y\rho x$ does not hold). Put

$$x \leqslant y \underset{\text{def}}{=} \boxed{x = y \text{ or } x\rho y.}$$

Show that $\leqslant$ is an order relation

5. If $\leqslant$ is an order relation and

$$x\rho y \underset{\text{def}}{=} \boxed{x \leqslant y \text{ and } x \neq y,}$$

then ρ is transitive and has the properties (*a*) and (*b*) of Exercise 4.

6. If φ is an isomorphism and a a greatest (or least, maximal or minimal) element, then $\varphi(a)$ is also a greatest (or least, maximal or minimal) element. Construct an example to show that this is not true for an arbitrary isotone mapping.

7. There are exactly two non-isomorphic posets of two elements.
8. In a finite poset there always exist a maximal as well as a minimal element.
9. An element which is at the same time minimal and maximal is not comparable with any element other than itself.
10. If a and b are maximal elements, then $\sup\{a, b\}$ exists iff $a = b$.
11. If a subset A of a poset P contains two distinct maximal elements of P, then $\sup A$ does not exist.
12. Construct an example to show that $\sup A$ may not exist even if A contains exactly one maximal element.
13. Formulate and prove the dual of Exercises 9–11.
14. Prove that

$$\inf_i (\sup_j \{x_{ij}\}) \geqslant \sup_j (\inf_i \{x_{ij}\}),$$

assuming that both sides of this inequality exist.
15. Define on some set P a relation ρ, which is not an equivalence and whose blocks form a partition.
16. Every poset is an ordered sum of one-element sets.
17. A poset is the cardinal (ordinal) sum of one-element sets iff it is trivial (is a chain).
18. If every summand of a cardinal (ordinal) sum decomposes into a cardinal (ordinal) sum, then we obtain a new expansion of the initial poset into a cardinal (ordinal) sum (this is called an *extension of the given expansion*).
19. Any two expansions of a poset into a cardinal (ordinal) sum have a common extension.
20. An ordinal sum is a chain iff every summand forms a chain.
21. A finite poset with zero and unity has always a composition series.
22. Construct an example of an infinite poset that has a composition series.

2

ORDINAL NUMBERS

In set-theoretic mathematics, an extensive use is made of transfinite induction. The possibility of its applications rests on the next assertion.

THEOREM 1. *The following properties of a poset P are equivalent* :

(i) (Minimum condition). *Every non-void subset of P is a poset, containing a minimal element.*
(ii) (Induction condition). *If all the minimal elements of P have some property $\mathcal{E}$ and if from the fact that all elements x in P, satisfying the condition $x < a$, have property $\mathcal{E}$ it follows that a too has property $\mathcal{E}$, then $\mathcal{E}$ is possessed by all elements of P.*
(iii) (Descending chain condition). *For every chain*

$$a_1 \geqslant a_2 \geqslant \ldots \geqslant a_k \geqslant \ldots$$

of elements in P, there is a number n such that

$$a_n = a_{n+1} = a_{n+2} = \ldots .$$

Proof. (i) $\Rightarrow$ (ii). We assume that the hypotheses of the induction condition are satisfied, and consider the set M of all elements in P, not having property $\mathcal{E}$. If the conclusion of hypothesis (ii) does not hold, then M is not void. In view of (i) there is a minimal element a in M. By hypothesis, a cannot be a minimal element of the poset P. If $x < a$, then $x \overline{\in} M$, and, consequently, x has property $\mathcal{E}$. But, then, a has property $\mathcal{E}$ by hypothesis. A contradiction.

(ii) ⇒ (iii). Let us agree to say that the element $a \in P$ has property $\mathcal{E}$, if every descending chain starting with the element a terminates, that is, satisfies condition (iii). Every minimal element $m \in P$ has property $\mathcal{E}$, since for the corresponding chain it is necessary to have

$$m = a_1 = a_2 = \dots$$

If $a \in P$ such that every $x < a$ has property $\mathcal{E}$, then we consider the chain

$$a = a_1 \geqslant a_2 \geqslant \dots .$$

If the equality sign does not occur everywhere, then there is an index i such that $a_1 = \dots = a_{i-1}$ and $a_{i-1} > a_i$. However, then the element a_i has property $\mathcal{E}$, that is, the chain

$$a_i \geqslant a_{i+1} \geqslant \dots ,$$

and hence the chain

$$a_1 \geqslant \dots \geqslant a_i \geqslant a_{i+1} \geqslant \dots$$

also terminates. Thus, the element a has property $\mathcal{E}$. In view of (ii) all elements in P have property $\mathcal{E}$, signifying also that P satisfies condition (iii).

(iii) ⇒ (i). Assume that a subset M of P is a poset, which does not have a minimal element. Choose an arbitrary $a_1 \in M$ and assume that the chain

$$a_1 > a_2 > \dots > a_n$$

is constructed from the elements in M. Since a_n is not minimal in M, there is an element $a_{n+1} < a_n$ in M. Thus, there is the infinite chain

$$a_1 > a_2 > \dots > a_n > \dots ,$$

which does not satisfy the condition (iii).

A chain, satisfying the minimum condition (and hence also the rest of the hypotheses of Theorem 1), is called a *well ordered set*. The elements of a well ordered set are called *ordinals* or *ordinal numbers*. Every finite chain is a well ordered set. A naturally ordered set of natural numbers is also well ordered. The set of all integers is not well ordered in their natural order, since it does not have a least element. However, it becomes well ordered, if the order is determined in the following way :

$$1 < 2 < 3 < \ldots < 0 < -1 < -2 < -3 < \ldots .$$

Another example of a chain that is not well ordered is the interval [0, 1], since the interval ($\frac{1}{2}$, 1), for instance, does not contain a minimal element.

From the definition of a well ordered set it follows that it always contains a least element 0. The succeeding elements are naturally denoted by 1, 2, . . . and so on. If α is some ordinal, then the lower cone α^{∇}, from which the element α is removed, is called an *initial segment* and denoted by $[0, \alpha)$. The symbol [0, 0) is understood as a void set. If $\alpha \neq 0$ and the initial segment $[0, \alpha)$ does not contain a greatest element, then the ordinal α is called a *limit*. An example of a limit ordinal is obtained by the number 0 in the well ordered set of integers considered above. For every α, among the ordinals of the upper cone α^{Δ}, different from α, there exists a least member, denoted by $\alpha + 1$.

THEOREM 2. *If α is a limit ordinal, then*

$$[0, \alpha) = \bigcup_{\beta<\alpha} [0, \beta).$$

Proof. If $\gamma \in [0, \alpha)$, then, since there is no element between γ and $\gamma + 1$, we have $\gamma + 1 \leqslant \alpha$. However, α being a limit number, equality is impossible. Thus,

$$\gamma \in [0, \gamma + 1) \subseteq \bigcup_{\beta<\alpha} [0, \beta),$$

that is,

$$[0, \alpha) \subseteq \bigcup_{\beta<\alpha} [0, \beta).$$

The reverse inclusion is obvious.

A chain C of a poset P is called a *maximal chain*, if for every element x in P, not belonging to C, the subset $C \cup x$ is not a chain.

THEOREM 3. *The following statements are equivalent* :

(i) (Axiom of choice). *For every non-void set $\mathfrak{M}$, there exists a mapping φ of the set $\mathfrak{P}(\mathfrak{M})$ into the set $\mathfrak{M}$ such that $\varphi(A) \in A$ for every $A \in \mathfrak{P}(\mathfrak{M})$, $A \neq \emptyset$.*

(ii) (Zermelo's Theorem). *On every non-void set, an order can be defined, which makes it into a well ordered set.*

(iii) (Hausdorff's Theorem). *Every chain in a poset can be extended to a maximal chain.*

(iv) (Kuratowski–Zorn Lemma). *If the upper cone of every chain of a poset P is not void, then P contains a maximal element.*

(v) *If every chain of a poset P has a least upper bound, then P contains a maximal element.*

Proof. (i) $\Rightarrow$ (v). Let P be a poset such that $\sup C$ exists for every chain of C in P and let φ be a mapping of the set $\mathfrak{P}(P)$ into the set P, such that $\varphi(X) \in X$ for $X \neq \emptyset$, which exists by the axiom of choice. Assume that P does not contain a maximal element. Then $\{x^{\Delta} \setminus x\}$ is a non-void set for every $x \in P$, allowing us to write

$$f(x) \underset{\text{def}}{=} \varphi(x^{\Delta} \setminus x).$$

Consider an element $a_0 \in P$. A subset H of P is said to have *Zorn's property*, provided that: (*a*) $a_0 \in H$; (*b*) if $x \in H$, then $f(x) \in H$; (*c*) if a chain C lies in H, then $\sup C \in H$. It is not hard to see that the intersection H_0 of all sets having Zorn's property has itself Zorn's property. Clearly the upper cone a_0^{Δ} has Zorn's property. Hence, $H_0 \subseteq a_0^{\Delta}$.

An element $a \in H_0$ is said to be *distinguished*, if the inequality $z < a$ for some $z \in H_0$ implies that $f(z) \leqslant a$.

LEMMA 1. *If a is a distinguished element, then the set*

$$B(a) = \{z \mid z \in H_0,\ z \leqslant a \text{ or } f(a) \leqslant z\}$$

has Zorn's property.

In fact, $a_0 \in B(a)$, since $a \in H_0 \subseteq a_0^{\Delta}$. Furthermore, if $x \in B(a)$, then we have $x < a$, $x = a$ or $f(a) \leqslant x$. In the first case we have $f(x) \leqslant a$, implying $f(x) \in B(a)$. If $x = a$, then we have $f(a) \leqslant f(x)$, which again implies that $f(x) \in B(a)$. In the third case we have $f(a) \leqslant x < f(x)$, whence it again follows that $f(x) \in B(a)$. Finally, if C is a chain and $C \subseteq B(a)$, then from $C \subseteq a^{\nabla}$ we have $\sup C \in a^{\nabla} \cap H_0 \subseteq B(a)$. However, if $c \nleqslant a$ for some $c \in C$, then $f(a) \leqslant c$, because $c \in B(a)$. But then $f(a) \leqslant \sup C$ and, consequently, $\sup C \in B(a)$.

LEMMA 2. *The set A of all distinguished elements has Zorn's property.*

In fact, a_0 is a distinguished element, since $z < a_0$ does not hold for any $z \in H_0$. Let $a \in A$ and $z < f(a)$ for some $z \in H_0$. By Lemma 1, we have $z \in H_0 \subseteq B(a)$. Consequently, $f(a) \leqslant z$ or $z \leqslant a$. However, the former inequality is impossible. In case the latter is satisfied, then for $z = a$ we have $f(z) = f(a)$ and for $z < a$, recalling that $a \in A$, we obtain

$$f(z) \leqslant a < f(a).$$

Thus, $f(a)$ is a distinguished element. Finally, let a chain C lie in A, $w = \sup C$ and let $z < w$ for some $z \in H_0$. If $z < c$ for some $c \in C$, then since c is a distinguished element, we have $f(z) \leqslant c \leqslant w$. If $z \nless c$, then we recall to mind the relation $z \in H_0 \subseteq B(c)$ that follows from Lemma 1. This implies that $z = c$ or $c < f(c) \leqslant z$. Thus, if $z \nless c$ for every $c \in C$, then $z \geqslant c$ for all $c \in C$. But, then, $z \geqslant w$, which is impossible. So, w is a distinguished element.

LEMMA 3. *H_0 is a chain.*

Indeed, all elements in H_0, by virtue of Lemma 2, are distinguished. If $a, b \in H_0$ then, according to Lemma 1, $b \in B(a)$. Consequently, $b \leqslant a$ or $a < f(a) \leqslant b$ holds.

We now note that by Lemma 3 and by hypothesis, there exists $w = \sup H_0$. However, then, from the definition of Zorn's property this implies that $w \in H_0$, $f(w) \in H_0$ and, consequently,

$$w < f(w) \leqslant w;$$

a contradiction.

(ii) ⇒ (iii). Let C be some chain in a poset P. If $C = P$, then there is nothing to prove. In the contrary case, consider a set $L = P \setminus C$ and assign to it an order $\lhd$, which makes it into a well ordered set. In general, an ordinal $\alpha \in L$ has property $\mathcal{E}$, if for all $\beta \lhd \alpha$, there exists a chain $C_\beta \subseteq L$, having the following properties: (*a*) $C \subseteq C_\beta$; (*b*) if $\gamma \lhd \beta \lhd \alpha$, then $C_\gamma \subseteq C_\beta$; (*c*) if $\beta \in C_\beta$, then β is not comparable with any $\xi \in C_\gamma$ for $\gamma \lhd \beta$. For the least ordinal $0 \in L$, we put

$$C_0 = \begin{cases} C, & \text{if } 0 \text{ is not comparable with any } c \in C\,; \\ C \cup 0, & \text{otherwise} \end{cases}$$

Thus, 0 has property $\mathcal{E}$. Assume that all ordinals smaller than α have property $\mathcal{E}$. In view of (*b*) the set $\overline{C} = \bigcup_{\beta \lhd \alpha} C_\beta$ is a chain. Put

$$C_\alpha = \begin{cases} \overline{C}, \text{ if } \alpha \text{ is not comparable with any } \bar{c} \in \overline{C}; \\ \overline{C} \cup \alpha \text{ otherwise.} \end{cases}$$

This shows, as it is easy to see, that α has property $\mathcal{E}$. Since, by Theorem 1, the set L satisfies the induction hypothesis, property $\mathcal{E}$ is possessed by all ordinals in L. In other words, the chains C_α exist for all $\alpha \in L$. We write

$$Q = \bigcup_{\alpha \in L} C_\alpha.$$

Clearly, Q is a chain. If it is not a maximal chain, then for some $\xi \overline{\in} Q$ the set $Q \cup \xi$ is also a chain. However, $\xi \overline{\in} C$ and is, consequently, some ordinal in L, where $\xi \overline{\in} C_\xi$. From condition (*c*) it follows that ξ is not comparable with some element in Q. This, however, contradicts the fact that $Q \cup \xi$ is a chain.

(iii) ⇒ (iv). Let P be a poset satisfying the premises of the Kuratowski–Zorn lemma. A one-element set a, where $a \in P$, is a chain. In view of (iii), this chain can be extended to a maximal chain C. By hypothesis, there is an element $c \in C^\Delta$. If c is not a maximal element of P, then $c < x$ for some $x \in P$. Of course, $x \in C^\Delta \setminus C$ and, consequently, $C \cup x$ is a chain, a contradiction to C being a maximal chain.

(iv) ⇒ (v). This is trivial.

(v) ⇒ (ii). Let $\mathfrak{M}$ be some non-void set. Consider all subsets of $\mathfrak{M}$ on which an order can be defined, making them into well ordered sets and denote by $\mathfrak{P}$ a collection of all such sets together with the corresponding order relations. It is clear that one and the same subset can fall in $\mathfrak{P}$ several times, but with different orders. The set $\mathfrak{P}$ is non-void, since it contains all one-element subsets. Define $\trianglelefteq$ on $\mathfrak{P}$, by putting

$$A \trianglelefteq B \underset{\text{def}}{=} \boxed{\begin{array}{c} A \subseteq B; \\ x \leqslant_A y \text{ implies } x \leqslant_B y \text{ for any } x, y \in A; \\ a <_B b, \text{ if } a \in A \text{ and } b \in B \setminus A. \end{array}}$$

It is easy to see that $\trianglelefteq$ is an order relation. If $\mathfrak{C}$ is any chain in $\mathfrak{P}$

and C the union of all subsets belonging to $\mathfrak{C}$, then define $\leqslant$ on C, by putting

$$x \leqslant y \underset{\text{def}}{=} \boxed{x \leqslant_A y \text{ for some } A \in \mathfrak{C}.}$$

Clearly, $\leqslant$ is a reflexive relation. If $x \leqslant y$ and $y \leqslant x$, then $x \leqslant_A y$ and $y \leqslant_B x$ for some $A, B \in \mathfrak{C}$. Since $\mathfrak{C}$ is a chain, we have, say, $A \trianglelefteq B$. Consequently, $x \leqslant_B y$ and $y \leqslant_B x$ hold, whence $x = y$. Hence, $\leqslant$ is an antisymmetric relation. If $x \leqslant y$ and $y \leqslant z$, then, as in the foregoing, we can assume that $x \leqslant_B y$ and $y \leqslant_B z$ for some $B \in \mathfrak{C}$. Consequently, $x \leqslant_B z$ and, hence, $x \leqslant z$. This shows $\leqslant$ to be an order relation. If $x, y \in C$, then $x \in A$ and $y \in B$ for some $A, B \in \mathfrak{C}$. However, we have, say, $A \trianglelefteq B$. Consequently, x and y are comparable in the sense of the order $\leqslant_B$, and, hence, in the sense of the order $\leqslant$. Thus, C is a chain. Finally, if there is a chain

$$x_1 \geqslant x_2 \geqslant \ldots,$$

then

$$x_1 \geqslant_{A_1} x_2 \geqslant_{A_2} x_3 \geqslant_{A_3} \ldots$$

for some $A_1, A_2, \ldots \in \mathfrak{C}$. Assume that for all $k < m$, it is established that $x_k \geqslant_{A_1} x_{k+1}$. Then we have $x_m \in A_1$ and $x_m \geqslant_{A_m} x_{m+1}$. If $A_m \trianglelefteq A_1$ then, evidently, $x_m \geqslant_{A_1} x_{m+1}$. Assuming $A_1 \trianglelefteq A_m$, if $x_{m+1} \overline{\in} A_1$, then $x_{m+1} >_{A_m} x_m$, which is impossible. However, if $x_{m+1} \in A_1$, and $x_{m+1} >_{A_1} x_m$, then we again obtain $x_{m+1} >_{A_m} x_m$. So, $x_m \geqslant_{A_1} x_{m+1}$. Thus, we get the chain

$$x_1 \geqslant_{A_1} x_2 \geqslant_{A_1} x_3 \geqslant_{A_1} \ldots$$

Since A_1 is a well ordered set, hence, by Theorem 1, we have

$$x_n = x_{n+1} = \cdots$$

for some n. This establishes that C is a well ordered set, that is, $C \in \mathfrak{P}$. If $A \in \mathfrak{C}$, then $A \subseteq C$ and $x \leqslant_A y$ implies $x \leqslant y$. If $x \in A$, and $y \in C \setminus A$, then $y \in B \setminus A$ for some $B \in \mathfrak{C}$. Of course, $A \trianglelefteq B$ and so, $x \leqslant_B y$, whence $x < y$. Thus, $A \trianglelefteq C$ for all $A \in \mathfrak{C}$, that is, $C \in \mathfrak{C}^\Delta$. If $D \in \mathfrak{C}^\Delta$, then $A \subseteq D$ for all $A \in \mathfrak{C}$ and, consequently, $C \subseteq D$. If x,

$y \in C$ and $x \leqslant_C y$, then $x \leqslant_A y$, for some $A \in \mathfrak{C}$ and, consequently, $x \leqslant_D y$. If $c \in C$ and $d \in D \setminus C$, then $c \in A$ for some $A \in \mathfrak{C}$ and, consequently, $c <_D d$. Thus, $C \trianglelefteq D$ and, consequently, $C = \sup \mathfrak{C}$. It is thereby established that the poset $\mathfrak{P}$ satisfies property (v). Thence, it follows that $\mathfrak{P}$ contains a maximal element Q. If $Q = \mathfrak{M}$, then everything is proved. In the contrary case take an element $q \mathrel{\overline{\in}} Q$ and place it after all elements of the set Q. It is obvious that the resulting well ordered set $\overline{Q}$ is an element of the poset $\mathfrak{P}$. Moreover, $Q \triangleleft \overline{Q}$, contradicting the fact that Q is a maximal element.

(ii) $\Rightarrow$ (i). If $\mathfrak{M}$ is some nonvoid set, then by (ii) we can take this to be well ordered. If A is a non-void subset of $\mathfrak{M}$, then denoting by $\varphi(A)$ a least element of A, we see that property (i) holds.

We emphasize that the motivation underlying Theorem 3 is to exhibit that the properties enumerated there are equivalent, not to prove the validity of any of them. The axiom of choice is in fact an axiom and cannot be derived from any other axiom of the set theory. *Throughout our subsequent exposition the axiom of choice is assumed to hold.* We shall dwell upon certain facts, whose proof involves the application of this axiom.

A subchain C' of C is called *cofinal,* if for every $x \in C$ there is an element $y \in C'$ such that $x \leqslant y$.

THEOREM 4. *Every chain contains a cofinal well ordered subset.*

Proof. Let $\mathfrak{P}$ be the set of all well ordered subsets of the given chain C. It is non-void, since it contains all one-element subsets. If A, $B \in \mathfrak{P}$, then put

$$A \trianglelefteq B \underset{\text{def}}{=} \boxed{\begin{array}{c} A \subseteq B \text{ and } b \in A^{\Delta} \text{ for every} \\ b \in B \setminus A. \end{array}}$$

It is easy to verify that $\trianglelefteq$ is an order relation. If

$$\mathfrak{C} : A_1 \trianglelefteq \ldots \trianglelefteq A_\alpha \trianglelefteq \ldots$$

is a chain in $\mathfrak{P}$, then we consider the union $A = \cup A_\alpha$. For every chain

$$a_1 \geqslant a_2 \geqslant \ldots$$

of elements in A, we have $a_i \in A_\alpha$, for an index α, such that $a_1 \in A_\alpha$. By Theorem 1 this is a terminating chain and further application of Theorem 1 shows that $A \in \mathfrak{P}$. Clearly, $A_\alpha \subseteq A$ for all α. If $x \in A \setminus A_\alpha$, then $x \in A_\beta \setminus A_\alpha$, where $A_\alpha \trianglelefteq A_\beta$, that is, $x \in A_\alpha^\Delta$. So, $A_\alpha \trianglelefteq A$. Hence, $A \in \mathfrak{C}^\Delta$. Therefore from the Kuratowski–Zorn lemma follows the existence of a maximal element $C' \in \mathfrak{P}$. Then C' is the desired cofinal subset. Indeed, if $x \in c$ and $x \in C'^\Delta$, then having placed the element x after all the elements of the set C', we obtain a well ordered set $\bar{C}$. Moreover, $C' \lhd \bar{C}$, in contradiction to C' being a maximal element. Consequently, $x \,\widetilde{\in}\, C'^\Delta$ and, hence, $x \leqslant y$ for some $y \in C'$.

We recall that the *direct product of sets* $\{P_\xi \mid \xi \in L\}$ is defined as the set of all functions a, assigning to every set P_ξ an element $a_\xi \in P_\xi$. The existence of such functions follows from the application of the axiom of choice to the union $\bigcup_{\xi \in L} P_\xi$. This way we obtain

THEOREM 5. *The direct product of any system of non-void sets is not void.*

Now, consider the direct product $\bar{P}$ of a family $\{P_\xi \mid \xi \in L\}$ of posets. We regard L also as a poset. The elements of $\bar{P}$ are represented in the form of a row (a_α). We define on $\bar{P}$ the relation $\lhd$, by putting

$$(a_\xi) \lhd (b_\xi) \underset{\text{def}}{=} \boxed{\begin{array}{l}\text{if } a_\alpha \nleqslant b_\alpha \text{ for some } \alpha,\\ \text{then } a_\beta < b_\beta \text{ for some } \beta < \alpha.\end{array}}$$

THEOREM 6. *If the poset L satisfies the minimum condition, then* $\trianglelefteq$ *on* $\bar{P}$*, defined by the condition*

$$a \trianglelefteq b \underset{\text{def}}{=} \boxed{a = b \text{ or } a \lhd b}$$

is an order relation.

Proof. The reflexivity of the relation $\trianglelefteq$ is obvious. The next lemma is immediate from the relevant definitions.

LEMMA. *If* $(a_\xi) \trianglelefteq (b_\xi)$ *and* μ *is a minimal element of the set L, then* $a_\mu \trianglelefteq b_\mu$.

Furthermore, if $(a_\alpha) \trianglelefteq (b_\alpha)$ and $(b_\alpha) \trianglelefteq (a_\alpha)$, then in view of the lemma, we have $a_\mu = b_\mu$ for every minimal element μ of L. Assume that the equality $a_\beta = b_\beta$ holds for all $\beta < \alpha$. If $a_\alpha \neq b_\alpha$ then $a_\alpha \nleq b_\alpha$ or $b_\alpha \nleq a_\alpha$ holds. Taking note of the inequalities $(a_\xi) \trianglelefteq (b_\xi)$ and $(b_\xi) \trianglelefteq (a_\xi)$, we arrive at the inequalities $a_\beta < b_\beta$ or $b_\beta < a_\beta$ for some $\beta < \alpha$, a contradiction. Consequently, $a_\alpha = b_\alpha$, whence, in view of Theorem 1, it follows that $a_\xi = b_\xi$ for all $\xi \in L$, that is, $(a_\xi) = (b_\xi)$. Hence, $\trianglelefteq$ is found to be an antisymmetric relation. Finally, assume that $(a_\xi) \trianglelefteq (b_\xi)$ and $(b_\xi) \trianglelefteq (c_\xi)$. If we have $(a_\xi) = (b_\xi)$ or $(b_\xi) = (c_\xi)$, then immediately it is clear that $(a_\xi) \trianglelefteq (c_\xi)$. However, if $(a_\xi) \triangleleft (b_\xi)$ and $(b_\xi) \triangleleft (c_\xi)$, but $(a_\xi) \ntrianglelefteq (c_\xi)$, then there is an index $\alpha \in L$, such that $a_\alpha \nleq c_\alpha$, and $a_\beta \nless c_\beta$ for all $\beta < \alpha$. Put $\alpha_1 = \alpha$ and assume that the chosen indices

$$\alpha_1 > \alpha_2 > \ldots > \alpha_n$$

are such that for all α_i $(i = 1, \ldots, n)$ there holds $a_{\alpha_i} \nleq b_{\alpha_i}$ or $b_{\alpha_i} \nleq c_{\alpha_i}$. If, say, $a_{\alpha_n} \nleq b_{\alpha_n}$, then for some $\beta < \alpha_n$ we must have $a_\beta < b_\beta$. By the choice of α we have $b_\beta \nleq c_\beta$ which allows us to put $\alpha_{n+1} = \beta$. Hence, we obtain an infinite sequence

$$\alpha_1 > \alpha_2 > \ldots$$

of elements in L, contradicting Theorem 1. It is thereby proved that $(a_\xi) \trianglelefteq (c_\xi)$, that is, the transitivity of the relation $\trianglelefteq$ is established.

A direct product $\bar{P}$ endowed with an order relation described in Theorem 6, is called an *ordered product* of posets P_ξ. If L is a trivial poset, we obtain the direct product, described in Chapter 1. The product is said to be ordered *lexicographically* if L is a well ordered set.

THEOREM 7. (Comparison Theorem). *For two well ordered sets P and P' there exists just one of the following possibilities*:

(i) *P is isomorphic to P'*;

(ii) *P is isomorphic to an initial segment of P'*;

(iii) *P' is isomorphic to an initial segment of P*.

Proof. Consider the well ordered set $\bar{P}$, obtained by adding some new element Ω to P, appearing after all the elements in P. Similarly by adding an element Ω' to P' we get a well ordered set $\bar{P'}$. Clearly, $P = [0, \Omega)$ and $P' = [0, \Omega')$.

LEMMA 1. *If θ is an isomorphism of a well ordered set Q into itself, then $\theta(x) \geqslant x$ for all $x \in Q$.*

Indeed, put $S = \{s \mid s \in Q, \theta(s) < s\}$. If the lemma is false, then S is a non-void set. If a is a least element of S, then $\theta(a) < a$ and, consequently, $\theta(a) \overline{\in} S$. Thence,

$$\theta(\theta(a)) \leqslant \theta(a) \leqslant \theta(\theta(a)),$$

that is, $\theta(a) = \theta(\theta(a))$ which, in view of $a \neq \theta(a)$, contradicts the fact that θ is a one-to-one mapping.

LEMMA 2. *A well ordered set cannot be isomorphic to any of its initial segments.*

In fact, if θ is an isomorphism of a well ordered set Q onto its initial segment $[0, a)$, then $\theta(a) < a$, violating Lemma 1.

LEMMA 3. *Let Q be a well ordered set, $a \in Q$ and φ be an isomorphism of Q onto an initial segment $[0, b')$ of a well ordered set Q'. If ψ is an isomorphism of an initial segment $[0, \alpha)$ onto an initial segment $[0, a')$ of the set Q', then $a' < b'$ and $\psi(x) = \varphi(x)$ for all $x \in [0, a)$.*

In fact, if $a' \geqslant b'$, then by applying successively the mappings ψ^{-1} and φ we obtain an isomorphism of the well ordered set $[0, a')$ onto an initial segment of itself, which is impossible by Lemma 2. However, if $\psi(x) \neq \varphi(x)$ for some $x < a$, then denoting by b the least element with these properties, we have $\psi(b) = \varphi(c)$, whence $b < c$. The successive application of φ and ψ^{-1} leads to an isomorphism of $[0, c)$ onto $[0, b)$ contradicting Lemma 2.

The stage is now set to conclude the proof of the theorem in hand. Consider the set S of all ordinals σ of the set $\bar{P}$, such that the segment $[0, \sigma)$ does not admit an isomorphism onto any initial segment of $\bar{P}'$. If $\Omega \overline{\in} S$, then either case (i) or (ii) holds. If this is not so, then S contains a least element α. Clearly, $\alpha \neq 0$. If $\alpha - 1$ exists, then there is an isomorphism φ of the segment $[0, \alpha - 1)$ onto the segment $[0, \beta')$ of $\bar{P}'$. If $\beta' = \Omega'$, then case (iii) holds. If not, then there exists an ordinal $\beta' + 1$. Hence, by putting

$$\bar{\varphi}(\xi) = \begin{cases} \varphi(\xi), & \text{if} \quad \xi < \alpha - 1, \\ \beta', & \text{if} \quad \xi = \alpha - 1, \end{cases}$$

we get an isomorphism $\bar{\varphi}$ of $[0, \alpha)$ onto $[0, \beta' + 1)$, contradicting the choice of α. Now we turn to the case when α is a limit ordinal. Then, according to Theorem 2,

$$[0, \alpha) = \bigcup_{\beta < \alpha} [0, \beta). \tag{*}$$

By the choice of α, for each β there is an isomorphism φ_β of $[0, \beta)$ onto the segment $[0, \beta')$ of $\bar{P}'$. If some β' equals Ω', then case (iii) holds. If not, then there is an element $\alpha' \in \bar{P}'$, the least among all those exceeding β'. If $\xi < \alpha$, then by (*) $\xi < \beta$ for some $\beta < \alpha$. We write

$$\varphi(\xi) = \varphi_\beta(\xi).$$

If $\xi < \gamma$, then we have, for example, $\gamma < \beta$. From Lemma 3, it follows that $\gamma' < \beta'$ and $\varphi_\gamma(\xi) = \varphi_\beta(\xi)$. Consequently, φ is a one-to-one mapping of the segment $[0, \alpha)$ into the segment $[0, \alpha')$. It is simple to verify that φ is a one-to-one and isotone mapping. If $\xi' \in [0, \alpha')$ then $\xi' \leqslant \beta'$ for some $\beta < \alpha$ and, consequently, for some $\xi < \beta$ we have

$$\xi' = \varphi_\beta(\xi) = \varphi(\xi).$$

Thus, φ is found to be an isomorphism of $[0, \alpha)$ onto the segment $[0, \alpha')$, contradicting the choice of α. It has thus been established that at least one of the cases (i)—(iii) holds. From Lemma 2, it is easy to derive that case (i) combines neither with the case (ii) nor with (iii). However, if cases (ii) and (iii) hold simultaneously, then the successive realization of isomorphisms indicated there allows us to obtain an isomorphism of the set P onto its initial segment, which again contradicts Lemma 2.

THEOREM 8. (Cantor–Bernstein). *If there is a one-to-one mapping of the set A into the set B and of B into A, then there exists a one-to-one mapping of A onto B.*

Proof. The successive realization of the postulated mappings gives a one-to-one map θ of the set A into itself. Let φ be the mapping of B into A referred to in the formulation. Put $A_0 = A$ and $A_1 = \varphi(B)$. Of course, we can assume that $A_1 \underset{\neq}{\subset} A_0$. Furthermore, put

$$A_i = \theta(A_{i-2}) \ (i = 2, 3, \ldots) \quad \text{and} \quad D = \bigcap_{i=1}^{\infty} A_i.$$

It is clear that

$$A_0 \supset A_1 \supset A_2 \supset \ldots,$$

$$A = D \cup [\bigcup_{i=0}^{\infty} (A_{2i} \setminus A_{2i+1})] \cup [\bigcup_{i=1}^{\infty} (A_{2i-1} \setminus A_{2i})],$$

$$A_1 = D \cup [\bigcup_{i=0}^{\infty} (A_{2i+2} \setminus A_{2i+3})] \cup [\bigcup_{i=1}^{\infty} (A_{2i-1} \setminus A_{2i})],$$

the union in the last two equalities being of mutually disjoint sets. Since $A_{2i} \setminus A_{2i+1}$ is mapped onto $A_{2i+2} \setminus A_{2i+3}$ by means of θ, it is easy to construct a one-to-one mapping ψ of A onto the set A_1. Hence the successive realization of the mappings ψ and φ^{-1} gives the desired mapping of A onto B.

The sets A and B are called *equivalent*, if there is a one-to-one mapping of A onto B.

THEOREM 9. (Theorem on the comparison of sets). *For any two sets A and B, there exists just one of the following possibilities*:

(i) *A is equivalent to B;*

(ii) *A is equivalent to a subset of B, but B is not equivalent to any subset of A;*

(iii) *B is equivalent to a subset of A, but A is not equivalent to any subset of B.*

Proof. From Theorem 7, in view of the possibility of well ordering of A and B, it follows that either one of the properties (i)–(ii) holds, or the hypotheses of Theorem 8 are satisfied. In the latter case, condition (i) is also satisfied. It is evident that the assertions made are mutually incompatible.

Theorem 9 serves as a basis for constructing the doctrine of *powers of sets*. In case (i) we say that the powers of sets A and B are equal, in case (ii) that the power of A is less than that of B, and in case (iii) that the power of B is less than that of A. We emphasize that these by no means define the notion of power itself but only assure the possibility of comparing the sets with respect to their powers. It is easy to show that if the power of A is less than that of B and the power of B is less than that of C, then the power of A is less than that of C. It is also simple to verify that for finite sets, a comparison with respect to powers is equivalent to a comparison with respect to number of elements.

Finally, we construct an auxiliary poset which will be needed in Chapter 5. With this objective, consider a set Q, which is the union of all direct products of finite numbers of copies of the set Z^+ of positive integers. The various rows

$$\mathfrak{m} = (m_1, \ldots, m_n),$$

where $n \geqslant 1$ and $m_i \in Z^+$ for all i can be regarded as the elements of Q. A passage from $\mathfrak{m}$ to the row

$$(m_1, \ldots, m_{i-1}, m_{i+1}, \ldots, m_n),$$

or to

$$(m_1, \ldots, m_{i-1}, h_1, \ldots, h_r, m_{i+1}, \ldots, m_n),$$

is called a *reduction*; where all $h_j < m_i$. A reduction of the former type is called a *rejection* and that of the latter type an *insertion*. Define on Q a relation $\leqslant$, putting $\mathfrak{m} \leqslant \mathfrak{n}$, if $\mathfrak{m} = \mathfrak{n}$ or if we can pass from $\mathfrak{n}$ to $\mathfrak{m}$ by a finite number of reductions. Clearly, $\leqslant$ is a reflexive and transitive relation. In order to prove that it is antisymmetric, consider a lexicographic product S of posets $Z_1, Z_2, \ldots$, each of which is a set of non-negative integers ordered in a natural way. In view of Theorem 6, S is a poset. A row $(z_1, z_2, \ldots)$ in S is called *special*, if $z_i \geqslant z_{i+1}$ for each $i = 1, 2, \ldots$ Each row $\mathfrak{m} \in Q$ corresponds to the special row $\theta(\mathfrak{m})$ obtained by writing out in non-increasing order all numbers encountered in $\mathfrak{m}$ and filling the remaining places by zeros. If the passage from $\mathfrak{n}$ to $\mathfrak{m}$ can be effected by means of a reduction, then both by rejection as well as insertion, we get $\theta(\mathfrak{m}) < \theta(\mathfrak{n})$. Hence, the relation $\mathfrak{m} \underset{\neq}{<} \mathfrak{n} \leqslant \mathfrak{m}$ implies an impossible relation $\theta(\mathfrak{m}) < \theta(\mathfrak{n}) \leqslant \theta(\mathfrak{m})$. Consequently, $\mathfrak{m} \leqslant \mathfrak{n} \leqslant \mathfrak{m}$ implies $\mathfrak{m} = \mathfrak{n}$, that is, Q is found to be a poset. This poset satisfies the minimum condition. Indeed, if this is not so, then by Theorem 1, there is an infinite chain

$$\mathfrak{m}_0 > \mathfrak{m}_1 > \mathfrak{m}_2 > \ldots$$

of rows in Q. We can assume that passage from $\mathfrak{m}_i$ to $\mathfrak{m}_{i+1}$ is realized by a single reduction. Hence, there arises an infinite chain

$$\theta(\mathfrak{m}_0) > \theta(\mathfrak{m}_1) > \theta(\mathfrak{m}_2) > \dots .$$

However, the existence of such a chain is, evidently, incompatible with the following assertion.

A set of special rows satisfies the minimum condition.

For proof, consider a non-void set A of special rows. Of these, we select a row $\mathfrak{a}_1$, having a least first coordinate, which is encountered the least number of times. Let $\mathfrak{a}_1 = (k_1, \dots, k_1, l_1, \dots)$, where $k_1 < l_1$. Let us choose in A a sequence of rows

$$\mathfrak{a}_1 = (k_1, \dots, k_1, \dots),$$

$$\mathfrak{a}_2 = (k_1, \dots, k_1, k_2, \dots, k_2, \dots)$$

$$\vdots \qquad \vdots$$

$$\mathfrak{a}_i = (k_1, \dots, k_1, k_2, \dots, k_2, \dots, k_i, \dots, k_i, \dots)$$

so that $k_1 > k_2 > \dots > k_i$ and among the rows in A, having the form

$$(\underbrace{k_1, \dots, k_1}_{m_1}, \underbrace{k_2, \dots, k_2}_{m_2}, \dots, \underbrace{k_{i-1}, \dots, k_{i-1}}_{m_{i-1}}, \dots),$$

the row $\mathfrak{a}_j$, $1 \leqslant j \leqslant i$, has a least $[(m_1 + \cdots + m_{j-1}) + 1]$-th coordinate, encountered a least number of times. By virtue of such a choice of elements, if $\mathfrak{b} \in A$ and $\mathfrak{b} < \mathfrak{a}_i$, then

$$\mathfrak{b} = (k_1, \dots, k_1, \dots, k_i, \dots, k_i, l, \dots),$$

where $l < k_i$. However, we can then find a row $\mathfrak{a}_{i+1}$, satisfying the requirements indicated above. It is clear that the construction described can be realized only a finite number of times. Of the elements obtained, the last one is also a minimal element of the set A.

Summing up above reasoning, we obtain

THEOREM 10. *The relation $\leqslant$ turns the set Q of finite rows $(m_1, \dots, m_n)$,* *where the m_i are positive integers, into a poset satisfying the minimum condition.*

Exercises

1. Formulate and prove a theorem dual to Theorem 1.
2. Any finite set satisfies the minimum and maximum conditions.
3. Prove that a poset with zero and unity, satisfying the minimum and maximum conditions, has a composition series. Construct an example showing that the converse is not true.
4. If C' is a cofinal subchain of C and there exists sup C' or sup C, then sup C = sup C'.
5. An ordered sum of well ordered sets $\{P_\alpha \mid \alpha \in L\}$ is well ordered iff L is a well ordered set.
6. An ordered sum of posets $\{P_\alpha \mid \alpha \in L\}$ satisfies the minimum condition iff it is satisfied by L and all P_α.
7. If an ordinal α precedes an infinite set of elements, then there is a limit ordinal $\alpha_0 \leqslant \alpha$ such that the interval $[\alpha_0, \alpha]$ contains only a finite number of elements.
8. If a chain C and its dual are both well ordered, then C is finite.
9. If a chain is not well ordered, then it contains a subchain dual to the natural series.
10. If a poset contains neither infinite chains nor infinite trivial posets, then it is finite.
11. Prove that the following properties of a poset P are equivalent: (*a*) P satisfies the minimum condition and does not contain infinite trivial posets; (*b*) any infinite subset of P does not satisfy the maximum condition.
12. Formulate the dual of the Kuratowski–Zorn lemma. Prove that it is equivalent to the axiom of choice.
13. Prove that the Kuratowski–Zorn lemma is equivalent to the corresponding statement formulated for well ordered chains.
14. Let L be a non-void set of subsets of some set $\mathfrak{M}$ and let the subset A of $\mathfrak{M}$ belong to L iff L contains all finite subsets of A. Prove that in the poset L, there exists a maximal element.
15. The ordered product of posets P_α satisfies the minimum condition iff this condition is satisfied by all P_α.
16. A lexicographic product of chains is a chain.
17. A lexicographic product of well ordered sets is well ordered.
18. Every poset can be represented in the form of a union of two disjoint subsets, of which one is well ordered and the other does not contain a least element.
19. Let T be a set of all trivially ordered subsets of a poset P. If $X, Y \in T$, then put $X \trianglelefteq Y$, if for every $x \in X$ there is $y \in Y$ such that $x \leqslant y$. Snow that $\trianglelefteq$ is an order relation and that the resultant poset T contains a greatest element, if P satisfies the hypotheses of the Kuratowski–Zorn lemma.

20. For two sets A and B, there exists just one of the following possibilities: (*a*) A is equivalent to B; (*b*) A is mapped onto B, but B is not mapped onto A; (*c*) B is mapped onto A, but A is not mapped onto B.

21. Show that the possibility (*b*) of Exercise 20 is realized iff the power of B is less than the power of A.

3

COMPLETE LATTICES

A poset is called a *complete lattice* when each of its non-void subsets has both a least upper bound (LUB) and a greatest lower bound (GLB) within the set. Examples of complete lattices are the interval [0, 1] with the usual order, the set of all subsets of some set, ordered by inclusion, and any finite chain. Clearly, every complete lattice must contain a least element 0 and a greatest element 1. Thus, for example, the set of all integers with the usual ordering is not a complete lattice.

THEOREM 1. *If a poset P has a unit element* 1 *and every non-void subset of P has a* LUB, *then P is a complete lattice.*

Proof. Let A be a non-void subset of P. The set A^{Δ} contains 1. So by hypothesis, there exists inf A^{Δ}. Applying Theorem 1.7, we deduce the existence of sup A, giving the desired proof.

From Theorem 1, it follows that the set of all subgroups of a given group, the set of all closed subsets of a topological space, every well ordered set with a greatest element etc., are complete lattices.

THEOREM 2. (Fixed-Point Theorem). *If* φ *be an isotone mapping of a complete lattice P into itself, then* $\varphi(a) = a$ *for some* $a \in P$.

Proof. Let S be a set of all elements s in P such that $s \leqslant \varphi(s)$.

Clearly, the zero element, which exists because P is a complete lattice, belongs to S, that is, S is not void. Consequently, there exists $a = \sup S$. Moreover, for any $s \in S$ we have $\varphi(a) \geqslant \varphi(s) \geqslant s$, whence $\varphi(a) \geqslant a$. Consequently, $\varphi[\varphi(a)] \geqslant \varphi(a)$, implying that $\varphi(a) \in S$ and, hence $a \geqslant \varphi(a)$. Thus, $a \geqslant \varphi(a) \geqslant a$, that is, $\varphi(a) = a$.

Theorem 2 does not admit a converse: the three-element poset depicted in Figure 2, is not a complete lattice, even though all of its isotone

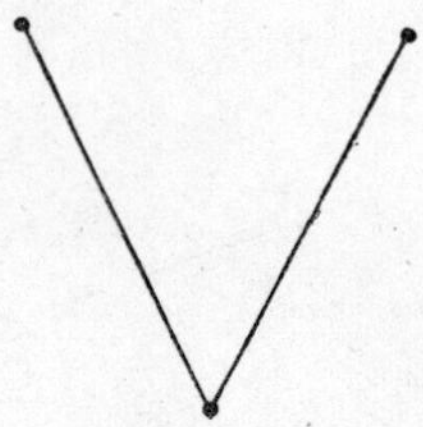

Figure 2.

mappings into itself have evidently a fixed point. However, we have

THEOREM 3. (Kogalovskii). *If each isotone mapping of a poset P into itself has a fixed point, then every maximal chain in P is a complete lattice.*

Proof. Suppose that P satisfies the hypotheses of the theorem. Then the following lemmas hold.

LEMMA 1. *If C is a chain in P, then $C^{\vartriangle}$ and $C^{\triangledown}$ are not void.*

Indeed, by Theorem 4 of Chapter 2, C contains a cofinal well ordered subset W. Let $x \in P$. If $W \subseteq x^{\triangledown}$, then for any $c \in C$ and for a suitable $w \in W$, we have

$$c \leqslant w \leqslant x,$$

that is, $x \in C^{\vartriangle}$. In the contrary case, $W \setminus x^{\triangledown}$ is a non-void set for every $x \in P$. Denote by $\varphi(x)$ the least element of this set. If $x, y \in P$ and $x \leqslant y$, then $x^{\triangledown} \subseteq y^{\triangledown}$. Hence

$$W \setminus x^{\triangledown} \supseteq W / y^{\triangledown}$$

and, consequently, $\varphi(x) \leqslant \varphi(y)$. This means that φ is an isotone mapping. Since $x \in x^{\triangledown}$ but $\varphi(x) \overline{\in} x^{\triangledown}$ for all $x \in P$, it follows that φ has no fixed point, contradicting the hypothesis. The proof of the second statement is immediate from our examination of the poset dually isomorphic to P.

Passing on to the proof of theorem, consider a maximal chain C. By Lemma 1, it contains a greatest and a least element, because from the maximality, it is easy to follow that $C^{\vartriangle}, C^{\triangledown} \subseteq C$. If C is not a complete lattice, then, by Theorem 1, it contains a non-void set V such that $\inf_C V$ does not exist. Let $U = C \cap V^{\triangledown}$. Clearly, the least element of C belongs to U, that is, the chain U is not void. By V^* we denote a chain, dual to V. An element $v \in V$, regarded as an element of V^*, is denoted by v^*. In a similar sense, we assign the symbol x^{**} to an element $x^* \in V^*$. By Theorem 2.4, the chains U and V^* contain cofinal well ordered subsets S and T^*, respectively.

LEMMA 2. *If* $x \in P$ *and* $S \subseteq x^{\triangledown}$, *then*, $T^* \nsubseteq (x^{\vartriangle})^*$.

In fact, if $S \subseteq x^{\triangledown}$ and $T^* \subseteq (x^{\vartriangle})^*$, then $T^{**} \subseteq x^{\vartriangle}$. If $v \in V$, then for some $t^* \in T^*$ we have $t^* \geqslant v^*$ and, consequently, $v \geqslant t^{**} \geqslant x$. If $z \in C$ and $z \leqslant v$ for all $v \in V$, then $z \in U$. For some $s \in S$ we have $z \leqslant s \leqslant x$. Hence, $x = \inf_C V$, in contradiction with the choice of V.

We return to the proof of the theorem. If $x \in P$, we put

$$\varphi(x) = \begin{cases} \inf_S (S \setminus x^{\triangledown}), & \text{if } S \nsubseteq x^{\triangledown}, \\ [\inf_{T^*} (T^* \setminus (x^{\vartriangle})^*)]^*, & \text{if } S \subseteq x^{\triangledown}. \end{cases}$$

In view of Lemma 2, φ is a mapping of the set P into itself. If $x, y \in P$ and $x \leqslant y$, then there holds either $S \nsubseteq x^{\triangledown}, y^{\triangledown}$ or $S \nsubseteq x^{\triangledown}$ and $S \subseteq y^{\triangledown}$, or $S \subseteq x^{\triangledown}$. In the first case, noting that $x^{\triangledown} \subseteq y^{\triangledown}$, we get

$$\varphi(x) = \inf_S (S \setminus x^{\triangledown}) \leqslant \inf_S (S \setminus y^{\triangledown}) = \varphi(y).$$

In the second case, we have

$$\varphi(x) = \inf_S (S \setminus x^{\triangledown}) \in S \subseteq U \subseteq V^{\triangledown},$$

and

$$\varphi(y) = (\inf_{T^*} (T^* \setminus (y^{\vartriangle})^*))^* \in T^{**} \subseteq V;$$

whence

$$\varphi(x) \leqslant \varphi(y).$$

In the third case, noticing that $x^{\Delta} \supseteq y^{\Delta}$, we infer

$$p^* = \inf_{T^*}(T^* \setminus (x^{\Delta})^*) \geqslant \inf_{T^*}(T^* \setminus (y^{\Delta})^*) = q^*;$$

whence

$$\varphi(x) = p^{**} \leqslant q^{**} = \varphi(y).$$

Hence, φ is an isotone mapping. By hypothesis, for some $a \in P$ the equality $a = \varphi(a)$ holds. If $S \nsubseteq a^{\nabla}$, then

$$a = \varphi(a) = \inf_S(S \setminus a^{\nabla}) \in S \setminus a^{\nabla}.$$

If $S \subseteq a^{\nabla}$, then

$$a = \varphi(a) = (\inf_{T^*}(T^* \setminus (a^{\Delta})^*))^* \in T^{**} \setminus a^{\Delta}.$$

In the former case, we get $a \overline{\in} a^{\nabla}$ and in the latter $a \overline{\in} a^{\Delta}$. However, either of them is impossible.

An isotone mapping φ of a poset P into itself is said to be a *closure operator*, if $\varphi(x) \geqslant x$, and $\varphi[\varphi(x)] = \varphi(x)$ for all $x \in P$. The examples of a closure operator are of a great variety. Thus, in the complete lattice of subspaces of a topological space, the closure operator is a mapping which assigns to every subspace its closure. If with each element of a complete lattice $\mathfrak{P}(\boldsymbol{L})$, where $\boldsymbol{L}$ is a linear space over a field, we associate its linear hull, then also we get a closure operator. In a poset P with element 1, the mapping $\varphi(x) = 1$ for all $x \in P$ is a closure operator.

If φ is a closure operator, then $\varphi(x)$ is called the *φ-closure* of x. An element, coinciding with its φ-closure, is called *φ-closed.* In the examples considered above, a closed subspace, a linear subspace and the unit element, constitute the φ-closed elements.

THEOREM 4. *If φ is a closure operator on a poset P, $A \subseteq P$ is a subset consisting of φ-closed elements and if $a = \inf A$ exists, then a is the φ-closed.*

Proof. Since $a \leqslant x$ for all $x \in A$, then $\varphi(a) \leqslant \varphi(x) = x$ for all $x \in A$ and, consequently, $\varphi(a) \leqslant a$; the reverse inequality follows from the definition of the closure operator.

THEOREM 5. *If φ is a closure operator on a complete lattice P, then the poset L of all φ-closed elements regarded as a subset of the poset P is also a complete lattice. In addition, for every non-void subset A of L we have*

$$\inf_L A = \inf_P A,$$

and

$$\sup_L A = \varphi(\sup_P A).$$

Proof. Let 1 be the unit element of the complete lattice P. Since $\varphi(1) \geqslant 1 \geqslant \varphi(1)$, it follows that 1 belongs to L and is, evidently, the unit element of this poset. Furthermore, if A is a non-void subset of L, then the element $a = \inf_P A$ is φ-closed by Theorem 4 and, of course, $a \leqslant x$ for all $x \in A$. If $v \in L$ and $v \leqslant x$ for all $v \in A$, then $v \leqslant a$. So, $a = \inf_L A$. Now Theorem 1 allows us to infer that L is a complete lattice. Further, let $b = \sup_P A$ and $\overline{b} = \sup_L A$. It is clear that $\overline{b} \in L$ and that $\overline{b} \geqslant b$, since $\overline{b} \geqslant x$ for all $x \in A$. Thence, $\overline{b} = \varphi(\overline{b}) \geqslant \varphi(b)$. The inequality $\overline{b} \leqslant \varphi(b)$ holds, because $\varphi(b) \geqslant \varphi(x) = x$ for all $x \in A$. So $\overline{b} = \varphi(b)$, giving the required proof.

THEOREM 6. *Let $P = \mathfrak{P}(M)$, where M is some poset. Then the mapping*

$$\varphi(A) \underset{\text{def}}{=} A^{\Delta\nabla}$$

is a closure operator on the complete lattice P.

Proof. It is immediate from property (i) of Theorem 1.6, that φ is an isotone mapping. The relation $A \subseteq A^{\Delta\nabla}$ is also contained in this theorem. The equality $A^{\Delta\nabla\Delta\nabla} = A^{\Delta\nabla}$ follows from property (iii) of Theorem 1.6.

Any poset can be embedded in a complete lattice.

THEOREM 7. *Let M be an arbitrary poset, let $P = \mathfrak{P}(M)$ and $L = \{A \mid A \in \mathfrak{P}(M), A^{\Delta\nabla} = A\}$. Then L is a complete lattice and there is a mapping θ of the set M into L, such that*

(i) θ *is an isomorphism of M onto a subset of the poset L;*

(ii) *if* $A \subseteq M$ *and* $\inf_M A$ *exist, then*

$$\theta(\inf_M A) = \inf_L \theta(A);$$

(iii) *if* $A \subseteq M$ *and* $\sup_M A$ *exist, then*

$$\theta(\sup_M A) = \sup_L \theta(A);$$

(iv) *If* $Z \in L$ *and* $Z \neq \phi, M$, *then there are subsets A and B of M such that* $Z = \sup_L \theta(A) = \inf_L \theta(B)$†.

Proof. Theorem 6 allows us to consider a closure operator $\Delta\nabla$ on a complete lattice P, and by Theorem 5 L as set of all $\Delta\nabla$-closed elements of a complete lattice P forms a complete lattice. If $x \in M$, then we put $\theta(x) = x^{\Delta\nabla}$ (here x is regarded as a one-element subset) and verify the validity of properties (i)–(iv). The relevant arguments are formulated in the form of a series of lemmas. For this, it should be kept in view that the order relation in the lattice L coincides with the inclusion relation $\subseteq$.

LEMMA 1. *If* $v \in M$, *then* $t \in v^{\Delta\nabla}$ *iff* $t \leqslant v$.

In fact, if $t \leqslant v$, then for every $x \in v^{\Delta}$ we have $t \leqslant v \leqslant x$, implying that $t \in v^{\Delta\nabla}$. If, conversely, $t \in v^{\Delta\nabla}$ then $t \leqslant x$ for all $x \in v^{\Delta}$. In particular, $t \leqslant v$, since $v \in v^{\Delta}$.

LEMMA 2. *If* $x, y \in M$, *then* $x \leqslant y$ *iff* $\theta(x) \leqslant \theta(y)$.

In fact, if $x \leqslant y$, then, in view of Lemma 1, either $x \in y^{\Delta\nabla}$ or $x \subseteq y^{\Delta\nabla}$ in P. But then $x^{\Delta\nabla} \subseteq y^{\Delta\nabla}$, that is, $\theta(x) \subseteq \theta(y)$. However, if $\theta(x) \subseteq \theta(y)$, then noticing Lemma 1, we get $x \in x^{\Delta\nabla} \subseteq y^{\Delta\nabla}$.

A further application of Lemma 1 gives $x \leqslant y$.

LEMMA 3. θ *is an isomorphism.*

In fact, if $\theta(x) = \theta(y)$, then, by Lemma 2, we have $x \geqslant y \geqslant x$, that

†The condition $Z \neq \phi, M$ can be omitted if we make the convention that $\sup_L \theta(\phi) = \phi$ and $\inf_L \theta(\phi) = M$.

is, $x = y$. Thus, θ is a one-to-one mapping. It was established in Lemma 2 that order is preserved.

LEMMA 4. *If* $a = \inf_M A$, *then* $\theta(a) = \inf_L \theta(A)$.

Proof. In view of Theorem 5, it will suffice to establish that $a^{\Delta\nabla} = \bigcap_{x \in A} x^{\Delta\nabla}$. But, by Lemma 1, $t \in a^{\Delta\nabla}$ implies that $t \leqslant a$. Hence, $t \leqslant x$ for all $x \in A$. According to Lemma 1, $t \in x^{\Delta\nabla}$ for all $x \in A$, that is, $t \in \bigcap_{x \in A} x^{\Delta\nabla}$, or $a^{\Delta\nabla} \subseteq \bigcap_{x \in A} x^{\Delta\nabla}$. By the same reasoning conducted in reverse order, we prove the reverse inclusion.

LEMMA 5. *If* $a = \sup_M A$, *then* $v \geqslant a$, *iff* $v \in (\bigcup_{x \in A} x^{\Delta\nabla})^{\Delta}$.

In fact, if $v \geqslant a$, then $v \geqslant x$ for all $x \in A$, whence, taking note of Theorem 1.6, we get $v \in x^{\Delta} = (x^{\Delta\nabla})^{\Delta}$. Since this holds for all $x \in A$, hence $v \in (\bigcup_{x \in A} x^{\Delta\nabla})^{\Delta}$. By the same reasoning conducted in a reverse order, we show that $v \in (\bigcup_{x \in A} x^{\Delta\nabla})^{\Delta}$ implies $v \geqslant x$.

LEMMA 6. *If* $a = \sup_M A$, *then* $\theta(a) = \sup_L \theta(A)$.

In fact, from Lemma 5, it follows that $a^{\Delta} = (\bigcup_{x \in A} x^{\Delta\nabla})^{\Delta}$ whence, taking note of Theorem 5, we get

$$\theta(a) = a^{\Delta\nabla} = (\bigcup_{x \in A} x^{\Delta\nabla})^{\Delta\nabla} = \sup_L \theta(A).$$

LEMMA 7. *If* $\phi \neq A \subseteq M$, *then*

$$A^{\Delta\nabla} = \sup_L \theta(A^{\Delta\nabla}) = \inf_L \theta(A^{\Delta}).$$

Proof. We first note that from Theorem 1.6 follow the relations

$$\bigcup_{x \in A^{\Delta\nabla}} x^{\Delta\nabla} \subseteq A^{\Delta\nabla\Delta\nabla} = A^{\Delta\nabla} \subseteq \bigcup_{x \in A^{\Delta\nabla}} x^{\Delta\nabla}.$$

Consequently, $A^{\Delta\nabla} = \bigcup_{x \in A^{\Delta\nabla}} x^{\Delta\nabla}$ and by Theorem 5 and Theorem 1.6, we have

$$\sup_L \theta(A^{\Delta\nabla}) = (\bigcup_{x \in A^{\Delta\nabla}} x^{\Delta\nabla})^{\Delta\nabla} = A^{\Delta\nabla\Delta\nabla} = A^{\Delta\nabla},$$

Further, if $t \in \bigcap_{x \in A^\Delta} x^{\Delta\nabla}$, then by Lemma 1, $t \leqslant x$ for all $x \in A^\Delta$, that is, $t \in A^{\Delta\nabla}$. However, if $t \in A^{\Delta\nabla}$ then by Lemma 1, $t \leqslant x$ for all $x \in A^\Delta$ and $t \in \bigcap_{x \in A^\Delta} x^{\Delta\nabla}$. Thus, noting Theorem 5, we get

$$\inf_L \theta(A^\Delta) = \bigcap_{x \in A^\Delta} x^{\Delta\nabla} = A^{\Delta\nabla}.$$

This completes the proof of Theorem 7.

The completion of a poset described in Theorem 7 is a generalization of the well known structure of real numbers as cuts or sections of a set of rational numbers and is called *completion by cuts* (cf. Exercise 11). The completion by cuts is, in a certain sense, the least among the completions of a given poset.

THEOREM 8. *Let M, L and θ have the same meaning as in Theorem 7. If φ is an isomorphism of the poset M into some complete lattice S, then there is an isomorphism ψ of the poset L into S, such that $\psi\theta(x) = \varphi(x)$ for all $x \in M$.*

Proof. Recall that the elements of the complete lattice L are the $\Delta\nabla$-closed elements of the complete lattice $P = \mathfrak{P}(M)$, that is, of a subset of M. Hence, for $X \in L$ we can write

$$\psi(X) \underset{\text{def}}{=} \sup_S \varphi(X).$$

Since $X \leqslant Y$ in L implies that $X \subseteq Y$, hence it is obvious that $X \leqslant Y$ implies $\psi(X) \leqslant \psi(Y)$. However, if $\psi(X) = \psi(Y)$ for some $X, Y \in L$, then for every $z \in Y^\Delta$ we have $\varphi(z) \geqslant \sup_S \varphi(Y) = \psi(Y)$. Hence, for all $x \in X$, we obtain

$$\varphi(x) \leqslant \psi(X) = \psi(Y) \leqslant \varphi(z).$$

Therefore $x \leqslant z$ for all $z \in Y^\Delta$, that is, $x \in Y^{\Delta\nabla} = Y$. Thus, $X \subseteq Y$. Similarly, the reverse inclusion is proved showing that ψ is an isomorphism of the poset L into the complete lattice S. From the definition it is immediate that

$$\psi\theta(x) = \sup_S \varphi(x^{\Delta\nabla})$$

holds for every $x \in M$. Further, if, $y \in x^{\Delta\nabla}$, then, by Lemma 1, established in the course of proving Theorem 7, we have $y \leqslant x$. Hence, $\varphi(y) \leqslant \varphi(x)$ for all $y \in x^{\Delta\nabla}$, that is,

$$\psi\theta(x) = \sup_S \varphi(x^{\Delta\nabla}) \leqslant \varphi(x).$$

From Lemma 1 referred above it also follows that $x \in x^{\Delta\nabla}$, whence

$$\varphi(x) \leqslant \sup_S \varphi(x^{\Delta\nabla}) = \psi\theta(x).$$

Thus, $\varphi(x) = \psi\theta(x)$, and the proof is complete.

It must be emphasized again that ψ is an isomorphism of a poset L but not of a complete lattice L, that is, for example, from $a = \sup_L A$ it does not follow, in general, that $\psi(a) = \sup_S \psi(A)$.

THEOREM 9. *Let M, L, and θ have the same meaning as in Theorem 7, and let S be a complete lattice, containing M as a poset* (that is, the order in M induces the order in S). *In order that there exists an isomorphism ψ of the complete lattice L onto S, such that $\psi\theta(x) = x$ for all $x \in M$, it is necessary and sufficient that S has the following properties*:

(i) *if $0 \neq s \in S$, then there is a subset A of M, such that $s = \sup_S A$;*

(ii) *if $1 \neq s \in S$, then there is a subset B of M, such that $s = \inf_S B$.*

Proof. The necessity of the properties indicated is immediate from Theorem 7. However, if these properties are possessed by S, then we denote by φ the natural embedding of M into S and consider the isomorphism ψ of the poset L into the complete lattice S, set forth in Theorem 8. If $s \in S$ and $s \neq 0{,}1$, then from properties (i)–(ii) it follows that

$$s = \sup_S A = \inf_S B$$

holds for some $A, B \subseteq M$. Clearly, $B \subseteq A^{\Delta}$. Hence, from Theorem 8 of present chapter and Theorems 5 and 7 of Chapter 1 it follows that

$$\begin{aligned} s = \sup_S \psi(\theta(A)) &\leqslant \psi(\sup_L \theta(A)) = \psi\,(\theta(\sup_M A)) \\ &= \psi\,(\theta(\inf_M A^{\Delta})) \leqslant \psi\,(\theta(\inf_M(B)) = \psi\,(\inf_L \theta(B)) \\ &\leqslant \inf_S \psi\,(\theta(B)) = s. \end{aligned}$$

Thus, $s = \psi(\sup_L \theta(A))$, that is, all elements s of S, other than 0 and 1, belong to $\psi(L)$. Since, by Theorem 1.5, for every such s we have

$$\psi(0) = \psi(\inf_L L) \leqslant \inf_S(\psi(L)) \leqslant s \leqslant \sup_S(\psi(L)) \leqslant \psi(\sup_L L) = \psi(1),$$

hence, $\psi(0) = 0$ and $\psi(1) = 1$. Consequently, ψ maps L onto S.

The restriction $s \neq 0$ and $s \neq 1$ in properties (i) and (ii) of Theorem 9 can be waived by putting $\sup \phi = 0$ and $\inf \phi = 1$.

Now, consider some non-void set $\mathfrak{M}$ and denote by Ξ the set of all equivalences defined on $\mathfrak{M}$. If $\rho, \sigma \in \Xi$, then set

$$\rho \leqslant \sigma \underset{\text{def}}{=} \boxed{\text{if } x\rho y, \text{ then } x\sigma y.}$$

It is easy to verify that the relation $\leqslant$ is reflexive, antisymmetric and transitive. Thus, Ξ becomes a poset.

THEOREM 10. *Ξ is a complete lattice.*

Proof. If R is a non-void subset of Ξ, then we define the relation θ by setting

$$x\theta y \underset{\text{def}}{=} \boxed{x\rho y \ \text{ for all } \rho \in R.}$$

It is simple to verify that θ is an equivalence. Clearly, $\theta \leqslant \rho$ for all $\rho \in R$. If $\sigma \leqslant \rho$ for all $\rho \in R$ and $x\sigma y$, then $x\rho y$ for all $\rho \in R$ and, consequently, $x\theta y$. Thus, $\sigma \leqslant \theta$. This shows that $\theta = \inf R$. It is routine to see that the equivalence I, defined by the condition: xIY for any $x, y \in \mathfrak{M}$, is the greatest element of the poset Ξ. It remains only to apply Theorem 1.

By way of an illustration of the application of Theorem 10, we state the next theorem.

THEOREM 11. *Every poset is an ordinal sum of its ordinally indecomposable subsets.*

Proof. If a poset P is expressed as an ordinal sum of its subsets $\{P_\alpha \mid \alpha \in L\}$, then the system of subsets $\{P_\alpha\}$ forms a partition. By

Theorem 1.2, this partition corresponds to the equivalence ρ, having the following properties:

(i) if $x\rho y$, $u\rho v$, $x \not\rho u$ and $x \leqslant u$, then $y \leqslant v$;

(ii) if $x \not\rho y$, then x and y are comparable.

Let R be the family of all equivalences on P, having properties (i)–(ii). In view of Theorem 10, there is an equivalence $\theta = \inf R$. If $x\theta y$, $u\theta v$, $x \not\theta u$ and $x \leqslant u$, then for some $\rho \in R$ we have $x\rho y$, $u\rho v$ and $x \not\rho u$, whence $y \leqslant v$. If $x \not\theta y$ then $x \not\rho y$ for some $\rho \in R$ and, consequently, x and y are comparable. Thus, the equivalence θ too has properties (i)–(ii). Let $\{Q_\alpha, \alpha \in T\}$ be a partition of P, corresponding to the equivalence θ, according to Theorem 1.1. If $\alpha, \beta \in T$, then we put

$$\alpha \trianglelefteq \beta \underset{\text{def}}{=} \boxed{\alpha = \beta \text{ or } x \leqslant y \text{ for some } x \in Q_\alpha, y \in Q_\beta.}$$

Clearly, $\trianglelefteq$ is a reflexive relation. If $\alpha \trianglelefteq \beta$ and $\beta \trianglelefteq \alpha$ but $\alpha \neq \beta$, then for some $x, y \in Q_\alpha$ and $u, v \in Q_\beta$ we have $x \leqslant u$ and $v \leqslant y$. In view of property (i) this implies that $u \leqslant x$. Consequently, $x = u$, contradicting $\alpha \neq \beta$. If $\alpha \trianglelefteq \beta$ and $\beta \trianglelefteq \gamma$ then, assuming that $\alpha \neq \beta$ and $\beta \neq \gamma$, we have $x \leqslant y$ and $u \leqslant v$ for some $x \in Q_\alpha$, $y, u \in Q_\beta$, $v \in Q_\gamma$. By property (i), $y \leqslant v$, hence $x \leqslant v$ and, consequently, $\alpha \trianglelefteq \gamma$. Hence, $\trianglelefteq$ is an order relation. From property (ii) it follows that T is a chain. If $x, y \in P$, then $x \in Q_\alpha$, $y \in Q_\beta$. Let $x \leqslant y$. If $\alpha = \beta$, then $x \leqslant_{Q_\alpha} y$. If $\alpha \neq \beta$, then $\alpha \lhd \beta$. Conversely, $x \leqslant_{Q_\alpha} y$ or $x \in Q_\alpha$, $y \in Q_\beta$ and $\alpha \lhd \beta$ imply that $x \leqslant y$. This shows that P is an ordinal sum of the subsets Q_α. If some Q_α is expanded into an ordinal sum of sets A and B, and σ is the corresponding equivalence on P, then $\sigma \leqslant \theta$. However, for $a \in A$ and $b \in B$ we have $a \,\theta\, b$ and $a \not\sigma b$. Consequently, $\sigma < \theta$, contradicting the choice of θ. This proves that the posets Q_α are ordinally indecomposable.

Exercises

1. If φ is an isomorphism of a poset P into a complete lattice L, then P is a complete lattice,

$$\varphi(\sup_P A) = \sup_L \varphi(A) \text{ and } \varphi(\inf_P A) = \inf_L \varphi(A)$$

for every non-void subset $A \subseteq P$.

2. An ordered sum of complete lattices $\{P_\alpha \mid \alpha \in L\}$, L a complete lattice, is a complete lattice.
3. The direct product of complete lattices is a complete lattice.
4. An ordered product of complete lattices $\{P_\alpha \mid \alpha \in L\}$, where L is a complete lattice, satisfying the minimum condition, is a complete lattice.
5. If φ is an isotone mapping of a complete lattice into itself, then the set of fixed points contains a least element.
6. If φ is a closure operator on a poset P, then all the maximal elements in P are φ-closed.
7. The mapping φ of a poset P into itself is a closure operator, iff for any $x, y \in P$, $\varphi(x) \geqslant x$ holds, and $x \leqslant \varphi(y)$ implies $\varphi(x) \leqslant \varphi(y)$.
8. If φ is a closure operator on a complete lattice L and Φ is a set of all φ-closed elements, then $\varphi(x) = \inf_L (\Phi \cap x^\Delta)$ for every $x \in L$.
9. If a subset U of a complete lattice L contains the unit element of L and $\inf_L A$ for every $A \subseteq U$, then the mapping $\varphi(x) = \inf_L (U \cap x^\Delta)$ is a closure operator on L, and U coincides with a set of all φ-closed elements.
10. Let Φ be the set of all closure operators on a poset P. If $\varphi, \psi \in \Phi$, then put $\varphi \trianglelefteq \psi$, if $\varphi(x) \leqslant \psi(x)$ for all $x \in P$. Verify that $\trianglelefteq$ is an order relation, making Φ a complete lattice.
11. The completion by cuts of a chain of rational numbers, lying between 0 and 1 is isomorphic to the segment [0, 1] with its natural ordering.
12. The completion by cuts of a chain is a chain.
13. The completion by cuts of a trivial poset P is isomorphic to a complete lattice, obtained by adjoining zero and unit element to P.
14. The completion by cuts of a direct product of two posets is isomorphic to a direct product of completion by cuts of the factors.
15. Let Ξ be a complete lattice of all equivalences over some set $\mathfrak{M}$. If $\Xi' \subseteq \Xi$ and $x, y \in \mathfrak{M}$, then put

$$x\theta y \underset{\text{def}}{=} \boxed{\begin{array}{l}\text{There exist a natural number } m,\\ \text{the elements } z_1, \ldots, z_{m-1} \in \mathfrak{M} \text{ and}\\ \text{the equivalences } \theta_1, \ldots, \theta_m \in \Xi'\\ \text{such that } x\theta_1 z_1,\ z_1\theta_2 z_2, \ldots,\\ z_{m-2}\theta_{m-1} z_{m-1},\ z_{m-1}\theta_m y.\end{array}}$$

Prove that

$$\theta = \sup_{\Xi} \Xi'.$$

4

LATTICES

A poset is called a *lattice*, if its every two-element subset has a GLB and a LUB. From Theorem 1.4, it is easy to deduce that in a lattice both bounds exist for any finite (non-void) subset. A complete lattice is, evidently, a lattice. Any chain is also a lattice. The set of integers (in their natural order) serves as an example of a lattice, which is not complete. A non-void subset P of a lattice L is called a *sublattice*, if $a, b \in P$ imply $c = \sup_L \{a, b\} \in P$ and $d = \inf_L \{a, b\} \in P$. As already noted, we then also have $c = \sup_P \{a, b\}$ and $d = \inf_P \{a, b\}$. From Theorem 3.7 it follows that every lattice is a sublattice of a complete lattice. It is simple to verify that any (closed) interval is a sublattice and, in particular, so is every one-element subset. A sublattice P is called *convex*, if $a, b \in P$, where $a \leqslant b$, imply $(a, b) \subseteq P$. The intersection of all sublattices of L, containing a given set $\mathfrak{M}$ is, evidently, a sublattice of which we say that it is *generated by the set* $\mathfrak{M}$. It is obvious that this is the least sublattice, containing $\mathfrak{M}$.

From the definition it is immediate that the poset dual to a lattice, is itself a lattice. It therefore follows that lattices admit the application of the duality principle.

On a lattice, we consider the following two operations:

$$a + b \underset{\text{def}}{=} \sup \{a, b\},$$

and

$$ab \underset{\text{def}}{=} \inf \{a, b\}.$$

Of these, the former is called *union* or *join* and the latter *intersection* or *meet*. From the definition and Theorem 1.4 it follows that in a lattice the operations of meet and join satisfy the following laws:

(i) $a + a = a$ $\quad aa = a$ (idempotent)

(ii) $a + b = b + a$ $\quad ab = ba$ (commutative)

(iii) $(a + b) + c = a + (b + c)$ $\quad (ab)\, c = a\, (bc)$ (associative)

(iv) $a\, (a + b) = a$ $\quad a + ab = a$ (absorptive).

From definition follows the next theorem too.

THEOREM 1. *The following properties of the elements a and b of a lattice L are equivalent*:

(*a*) $a \leqslant b$;

(*b*) $a + b = b$;

(*c*) $ab = a$.

Hence, for the zero and unit element, if they exist, the equalities

$$a0 = 0,\ a + 0 = a,\ a1 = a \text{ and } a + 1 = 1$$

hold for every $a \in L$.

Note also that (i) is a consequence of (iv), since the latter yields

$$a + a = a + a\, (a + a) = a,$$

and

$$aa = a\, (a + aa) = a.$$

We can arrive at the definition of a lattice, also by circumventing the notion of a partially ordered set.

THEOREM 2. *Let L be a set with a binary operation $+$ and $\cdot$, having* *properties* (ii)–(iv). *Then the relation*

$$a \leqslant b \underset{\text{def}}{=} \boxed{a + b = b}$$

is an order on L, and the resulting poset is found to be a lattice; moreover,

$$a + b = \sup \{a, b\},$$

and

$$ab = \inf \{a, b\}.$$

Proof. The reflexivity of the relation $\leqslant$ follows from property (i), which, as noted is a consequence of (iv). If $a \leqslant b$ and $b \leqslant a$, that is, $a + b = b$ and $b + a = a$, then in view of (ii), $a = b$, that is, the relation $\leqslant$ is antisymmetric. If $a + b = b$ and $b + c = c$, then, by applying (iii), we get

$$a + c = a + (b + c) = (a + b) + c = b + c = c,$$

showing transitivity of the relation $\leqslant$. Furthermore, taking note of (iii), (i) and (ii), we get

$$a + (a + b) = (a + a) + b = a + b,$$

and

$$b + (a + b) = b + (b + a) = b + a = a + b.$$

Consequently, $a \leqslant a + b$ and $b \leqslant a + b$. If $a \leqslant x$ and $b \leqslant x$, then, by using (i)–(iii) we have

$$(a + b) + x = a + (b + x) = a + x = x,$$

that is, $a + b \leqslant x$. We recall the definition of a LUB to obtain

$$a + b = \sup \{a, b\}.$$

Finally, from (ii) and (iv) it follows that $ab \leqslant a$ and $ab \leqslant b$. If $y \leqslant a$ and $y \leqslant b$, then, with the aid of (iii) and (iv), we obtain

$$y\,(ab) = [y\,(y + a)]\,b = yb = y\,(y + b) = y.$$

Hence, in view of (ii) and (iv)

$$y + ab = y\,(ab) + ab = ab,$$

that is, $y \leqslant ab$. Hence,

$$ab = \inf\{a, b\}.$$

Note that reasoning dually we are able to arrive at the same conclusion, if the relation $\leqslant$ is defined by the condition

$$a \leqslant b \underset{\text{def}}{=} \boxed{ab = a}.$$

From Theorem 1, if follows that lattices form a variety of algebras with two binary operations.

It is routine to see that properties (i)–(iv) appearing on the left-hand side are the dual of the respective properties on the right-hand side. Hence, by induction on the number of applications of these properties, it can be established that in an arbitrary lattice, the validity of any property, expressed in terms of lattice operations, implies that a dual property also holds.

THEOREM 3. *If $a \leqslant c$ and $b \leqslant d$, then $a + b \leqslant c + d$ and $ab \leqslant cd$.*

Proof. In view of Theorem 1, from the hypothesis it follows that $a + c = c$, $b + d = d$, $ac = a$ and $bd = b$. Hence

$$(a + b) + (c + d) = (a + c) + (b + d) = c + d,$$

and

$$(ab)\,(cd) = (ac)\,(bd) = ab,$$

whence also follow the desired inequalities.

A mapping φ of a lattice L into a lattice L' is called a *homomorphism* (or '*lattice-morphism*') if $\varphi\,(a + b) = \varphi\,(a) + \varphi\,(b)$ and $\varphi\,(ab) = \varphi\,(a)\,\varphi\,(b)$ for all $a, b \in L$. With the aid of Theorem 1, it is easy to deduce that every homomorphism is an isotone mapping. The converse is not true, as is shown in the example depicted in Figure 3. A homomorphic one-to-one correspondence is called an *isomorphism*.

$$\varphi(0) = 0, \quad \varphi(a) = \varphi(b) = c,$$
$$\varphi(1) = 1.$$

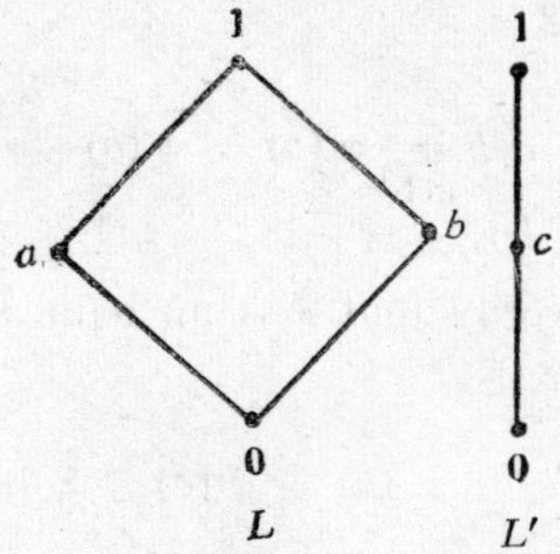

Figure 3.

φ is an isotone mapping, but

$$\varphi(a + b) = \varphi(1) = 1 \neq c + c = \varphi(a) + \varphi(b).$$

THEOREM 4. *If φ is an isomorphism of a poset L onto a poset L' and if L is a lattice, then L' is also a lattice and φ is an isomorphism of the lattices.*

Proof. If $a', b' \in L'$, then $a' = \varphi(a)$ and $b' = \varphi(b)$, where $a, b \in L$. But then, $\varphi(a + b) \geqslant a', b'$. If $v' \geqslant a', b'$, then $v \geqslant a, b$, where $v' = \varphi(v)$. Therefore, $v \geqslant a + b$ and, so, $v' \geqslant \varphi(a + b)$. Thus, $\varphi(a + b) = \varphi(a) + \varphi(b)$. The equality $\varphi(ab) = \varphi(a)\varphi(b)$ is proved by duality.

An equivalence $\equiv$ defined on a lattice L is called a *congruence* relation, if $a \equiv c$ and $b \equiv d$ imply that $a + b \equiv c + d$ and $ab \equiv cd$. Let $\equiv$ be a congruence relation on a lattice L. If $a \equiv b$, then $a + b \equiv a + a = a$ and $ab \equiv aa = a$, that is, each residue class of the congruence forms a sublattice. If, in addition, we have $a \leqslant x \leqslant b$, then $x = bx \equiv ax = a$. Consequently, each residue class of the congruence is a convex sublattice. Further, consider the quotient set $\bar{L} = L/\equiv$. Denoting by $\bar{a}$ the residue class of the congruence $\equiv$, defined by the element a, and putting

$$\bar{a} + \bar{b} \underset{\text{def}}{=} \overline{a + b},$$

and

$$\bar{a}\bar{b} \underset{\text{def}}{=} \overline{ab},$$

it is easy to see that the operations are well defined and satisfy properties (ii)–(iv). Hence, the quotient set $\overline{L}$, turned into a lattice, is said to be a *quotient lattice* of L with respect to the congruence $\equiv$. Clearly, the natural mapping is a homomorphism, also called *natural.*

Now, let φ be a homomorphism of a lattice L into a lattice L'. Define on L a relation ρ, putting

$$a\rho b \underset{\text{def}}{=} \boxed{\varphi(a) = \varphi(b)}.$$

It is a simple matter to verify that ρ is an equivalence relation. If $a\rho c$ and $b\rho d$, then

$$\varphi(a + b) = \varphi(a) + \varphi(b) = \varphi(c) + \varphi(d) = \varphi(c + d),$$

and

$$\varphi(ab) = \varphi(a)\,\varphi(b) = \varphi(c)\,\varphi(d) = \varphi(cd).$$

Consequently, $(a + b)\,\rho\,(c + d)$ and $(ab)\,\rho\,(cd)$, that is, ρ forms a congruence relation. This congruence is called the *kernel of the homomorphism* φ.

THEOREM 5. (Homomorphism Theorem). *Let φ be a homomorphism of a lattice L onto a lattice L' and let ρ be the kernel of this homomorphism. Then, there exists an isomorphism ψ of L' onto a quotient lattice L/ρ, such that $\psi\varphi(a) = \overline{a}$ for all $a \in L$, where $\overline{a}$ denotes the image of an element a under the natural homomorphism determined by the congruence ρ.*

Proof. If $a' \in L'$, then put $\psi(a') = \overline{a}$, where $\varphi(a) = a'$. If $\varphi(b) = a'$ for some $b \in L$, then $a\rho b$ by definition of the kernel of a homomorphism and, consequently, $\overline{a} = \overline{b}$. So the mapping ψ is well defined. If $a', b' \in L'$, then $a' + b' = \varphi(a + b)$, where $a' = \varphi(a)$ and $b' = \varphi(b)$. Hence,

$$\psi(a' + b') = \overline{a + b} = \overline{a} + \overline{b} = \psi(a') + \psi(b').$$

Similarly it is verified that

$$\psi(a'b') = \psi(a')\,\psi(b').$$

If $\psi(a') = \psi(b')$ and $a' = \varphi(a)$, $b' = \varphi(b)$, then $\overline{a} = \overline{b}$, that is, $a\rho b$,

Hence,

$$a' = \varphi(a) = \varphi(b) = b'.$$

Thus, ψ is an isomorphism. However, the equality $\psi\varphi(a) = \overline{a}$ holds by the definition of ψ.

In the case of groups and rings, every congruence is uniquely defined by any one of its residue classes. Unfortunately, a lattice, in general, does not possess such a property. For example, a congruence relation on a three-element chain, one of whose classes contains only zero, and the other both the remaining elements, is evidently different from the diagonal. However, zero in both the cases is a residue class.

A lattice L is called *relatively complemented*, if for every element c in any interval $[a, b]$ there is an element d, such that $c + d = b$ and $cd = a$. This element d is called a *relative complement of c in the interval* $[a, b]$. A complement is not defined uniquely. For example, in the lattice depicted in Figure 4, the complements of b relative to $[s, 1]$ are the

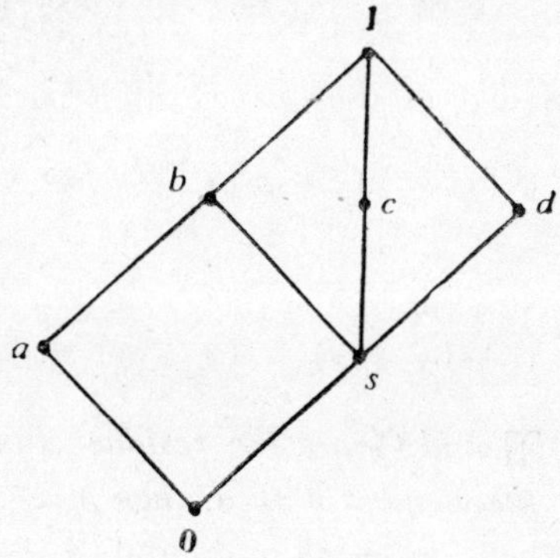

Figure 4.

elements c and d. However, each of the elements a and s serve as the unique complement of the other relative to the interval $[0, b]$. A lattice with zero and unit is called *complemented*, if all its elements have complements relative to $[0, 1]$. The complements in the interval $[0, 1]$ are simply called *complements*.

THEOREM 6. (Hashimoto). *Every congruence on a relatively complemented lattice L is defined by any of its residue classes.*

Proof. Let $\equiv$ and $\approxeq$ be two congruences on L and let $\mathfrak{R}$ be a residue class of $\equiv$, such that $a \approxeq b$, for every $a, b \in \mathfrak{R}$.

LEMMA 1. *If $x, y \geqslant a$ for some $a \in \mathfrak{R}$ and $x \equiv y$, then $x \approxeq y$.*

For the proof, denote by z a complement of the element xy relative to $[a, x + y]$. Then

$$z = z\,(x + y) \equiv zx \equiv zxy = a,$$

whence, $z \in \mathfrak{R}$. Hence,

$$x + y = z + xy \approxeq a + xy = xy$$

and, consequently,

$$x = x\,(x + y) \approxeq x\,(xy) = xy = (xy)\,y \approxeq (x + y)\,y = y.$$

LEMMA 2. *If $x, y \leqslant a$ for some $a \in \mathfrak{R}$ and $x \equiv y$, then $x \approxeq y$.*

In fact, denote by z a complement of the element $x + y$ relative to $[xy, a]$, to obtain

$$z = z + xy \equiv z + x \equiv z + (x + y) = a,$$

implying $z \in \mathfrak{R}$. Hence,

$$xy = z\,(x + y) \approxeq a\,(x + y) = x + y$$

and, consequently,

$$x = x + xy \approxeq x + (x + y) = (x + y) + y \approxeq xy + y = y.$$

LEMMA 3. *If $a \in \mathfrak{R}$ and $\mathfrak{C}$ are the residue classes of the congruence relation $\equiv$ defined by the element $b \leqslant a$, then $p \approxeq q$ for all $p, q \in \mathfrak{C}$.*

In fact, if $p \in \mathfrak{C}$, then $ap \equiv ab = b$ and $a + p \equiv a + b = a$. In view of Lemmas 1 and 2, we have $ap \approxeq b$ and $a + p \approxeq a$, whence

$$p = (a + p)\,p \approxeq ap \approxeq b.$$

Now to complete the proof we need only invoke the transitivity of the relation $\approxeq$.

LEMMA 4. If $x \equiv y$, then $x \approxeq y$.

For proof, we select an element $a \in \mathfrak{R}$ and consider the residue class $\mathfrak{C}$ of the congruence $\equiv$, defined by the element axy. In view of Lemma 3, the residue classes of the congruences $\equiv$ and $\approxeq$, defined by the element axy, coincide. Since, $x, y \geqslant axy$, hence, by Lemma 1, $x \approxeq y$.

The proof of Theorem 6 is immediate from Lemma 4.

We describe one more congruence property of relatively complemented lattices. The congruences $\equiv$ and $\approxeq$ on a lattice L are called *permutable*, if the relations $a \equiv x$ and $x \approxeq b$, where $a, b, x \in L$, imply the existence of an element $y \in L$, such that $a \approxeq y$ and $y \equiv b$. It is easy to verify that both the diagonal and the unit congruences commute with any congruence.

THEOREM 7. (Dilworth). *Any two congruences on a relatively complemented lattice L commute.*

Proof. Let $\equiv$ and $\approxeq$ be congruences on L, let $a, b, x \in L$, $a \equiv x$ and $x \approxeq b$. We preface the proof by the next lemma.

LEMMA. *If $a \leqslant x \leqslant b$, then for some $y \in [a, b]$, $a \approxeq y$ and $y \equiv b$ hold.* In fact, let y be a complement of x relative to $[a, b]$. Then,

$$y = by \approxeq xy = a,$$

and

$$y = a + y \equiv x + y = b.$$

Now, waiving the restrictions made in the lemma, we have

$$\begin{aligned} a = a + a \equiv a + x = a + x + x \approxeq a + b + x \\ \equiv x + b + x = b + x \approxeq b + b = b. \end{aligned}$$

Applying the lemma, with regard to the equivalences $\equiv$ and $\approxeq$ we obtain

$$a \approxeq u \equiv a + b + x,$$

and

$$a + b + x \approxeq v \equiv b,$$

where $a \leqslant u \leqslant a + b + x$ and $b \leqslant v \leqslant a + b + x$. Hence,

$$u = u\,(a + b + x) \approxeq uv \equiv (a + b + x)\,v = v$$

and, consequently, $a \approxeq uv$ and $uv \equiv b$.

A non-void subset I of a lattice L is called an *ideal*, if $x, y \in I$ imply

$x + y \in I$ and $t \in I$ for all $t \leqslant x$. Dually, we introduce the notion of a *filter*. If $x \in L$, then $x^{\triangledown}$ and $x^{\vartriangle}$ are an ideal and a filter, respectively, called *principal*. If a congruence θ on L is such that the quotient lattice L/θ has a zero, then it is easily verified that the complete pre-image of this zero is an ideal, called the *kernel ideal* of the congruence θ. Dually, we define the *kernel filter*. Of course, the zero of the lattice and its unit, if they exist belong to every ideal and filter, respectively. A kernel ideal (filter) which is the kernel of a homomorphism φ is called the kernel ideal (filter) of this homomorphism. Unlike the case of groups and rings, not every ideal of a lattice is a kernel. For example, the set $I = \{0, a, b, s\}$ of the elements of the lattice depicted in Figure 4, is an ideal. If this were the kernel ideal of a congruence $\equiv$, we would have the relations $c \equiv b + c \equiv 1$ and $d \equiv b + d \equiv 1$. But then, $0 \equiv cd \equiv 1$, which implies that $1 \in I$.

THEOREM 8. *If S is a filter, and I an ideal of the lattice L then the intersection $S \cap I$ is a convex sublattice of L. Every convex sublattice can be represented in the form of such an intersection. If a filter and an ideal are kernels, then their intersection is a residue class of some congruence relation.*

Proof. It is clear that the intersection $S \cap I$ is a sublattice. If $a \leqslant x \leqslant b$ and $a, b \in S \cap I$, then $x \in S$, since $a \in S$, and $x \in I$, because $b \in I$. Conversely, if V is a convex sublattice, then we consider a set $S = \bigcup_{v \in V} v^{\vartriangle}$. It is clear that $s \in S$ implies $x \in S$ for all $x \geqslant s$. If $s', s'' \in S$, then $s' \geqslant v'$ and $s'' \geqslant v''$, where $v', v'' \in V$. However, $s's'' \geqslant v'v'' \in V$, that is, $s's'' \in S$. Consequently, S is a filter. By duality it is verified that the set $I = \bigcup_{v \in V} v^{\triangledown}$ is an ideal. It is clear that $V \subseteq S \cap I$. However, if $x \in S \cap I$, then for a suitable $v', v'' \in V$ we have $v' \leqslant x \leqslant v''$.

Since the sublattice V is convex, it follows that $x \in V$. So $S \cap I \subseteq V$. Finally, let I and S be the kernel ideal and filter of congruences ρ and σ, respectively. Consider the direct product H of the quotient lattices L/ρ and L/σ. Let χ be the homomorphism of the lattice L into this direct product, defined by the equality $\chi(x) = [\varphi(x), \psi(x)]$, where φ and ψ are the natural homomorphisms. It is easily verified that $v \in V = I \cap S$ iff $\chi(v) = (0, 1)$. Hence, V forms a residue class of the kernel of the homomorphism χ.

An ideal S in a lattice L is called *standard*, if for any $a, b \in L$ and $s \in S$

the inequality $a \leqslant b + s$ implies $a = x + t$, where $x \leqslant b$, $t \in S$. It is easy to see that the zero ideal is standard. A standard ideal is itself a lattice. In the lattice depicted in Figure 4, the ideal s^{∇} is standard but the ideal a^{∇} is not so.

THEOREM 9. (Grätzer–Schmidt). *Every standard ideal S in a lattice L is the kernel ideal of the congruence* θ, *defined by the condition*

$$x\theta y \underset{\text{def}}{=} \boxed{x + y = xy + s \text{ for some } s \in S.}$$

Proof. We start with some preparatory lemmas.

LEMMA 1. *If* $s \in S$, *then for any* $x, y \in L$, *there is an element* $t \in S$, *such that* $x(y + s) = xy + t$.

In fact, since $x(y + s) \leqslant y + s$, it follows that $x(y + s) = b + t$, where $b \leqslant y$, $t \in s$. But

$$b \leqslant x(y + s) \leqslant x$$

and, consequently, $b \leqslant xy$. Hence

$$x(y + s) = x(y + s) + xy = b + t + xy = xy + t.$$

LEMMA 2. *If* $x + y = a + s$, *where* $x, y \in L$, $a \leqslant xy$, $s \in S$, *then* $x\theta y$.

Indeed, we have

$$x + y = xy + x + y = xy + a + s = xy + s$$

and this proves the assertion.

LEMMA 3. *If* $x\theta y$, $y\theta z$ *and* $y \leqslant x + z$, *then* $x\theta z$.

Indeed, from the definition we have

$$x + y = xy + s',$$

and

$$y + z = yz + s'',$$

where $s', s'' \in S$. In view of Lemma 1, there is $t', t'' \in S$, such that

$$xy = xy(y + z) = xy(yz + s'') = xyz + t',$$

and

$$yz = yz\,(x + y) = yz\,(xy + s') = xyz + t''.$$

Thus,

$$x + z = x + y + z = xy + yz + s' + s''$$
$$= xyz + (t' + t'' + s' + s''),$$

whence

$$x + z = xz + xyz + (t' + t'' + s' + s'') = xz + (t' + t'' + s' + s'')$$

and, consequently, $x\theta z$.

LEMMA 4. *If $x\theta u$ and $y\theta v$, then $(x + y)\,\theta\,(u + v)$ and $xy\theta uv$.*
In fact, from the definition of relation θ, we obtain

$$x + u = xu + s',$$

and

$$y + v = yv + s''$$

for suitable $s', s'' \in S$. We thereby deduce that

$$(x + y) + (u + v) = (xu + yv) + s' + s''.$$

Since

$$xu + yv \leqslant (x + y)\,(u + v) + (x + y)\,(u + v) = (x + y)\,(u + v),$$

by Lemma 2 it follows that $(x + y)\,\theta\,(u + v)$. Furthermore, applying Lemma 1, we get

$$xy + uv = (xy + uv)\,(x + u)\,(y + v)$$
$$= (xy + uv)\,(xu + s')\,(yv + s'')$$
$$= (xy + uv)\,[(xu + s')\,yv + \bar{\bar{s}}]$$
$$= (xy + uv)\,(xuyv + \bar{s} + \bar{\bar{s}}) = xuyv + s,$$

where $\bar{s}, \bar{\bar{s}}, s \in S$. Consequently, $xy\theta uv$.

LEMMA 5. *θ is a congruence.*

Since the reflexivity and symmetry of θ are obvious, it is sufficient in view of Lemma 4 to establish the transitivity of this relation. If $x\theta y$ and $y\theta z$, then from Lemma 4 it follows that

$$x\theta y\theta\,(x + y + z),$$

which implies by Lemma 3 that $x\theta\,(x + y + z)$. We apply again Lemma 4, to obtain

$$x = x\,(x + z)\,\theta\,(x + y + z)\,(x + z) = x + z.$$

In view of x and z being equivalent, we have

$$x\theta\,(x + z)\,\theta z,$$

and then the relation $x\theta z$ follows from Lemma 3.

To prove the theorem in hand, it remains to establish that S coincides with the kernel ideal of the congruence θ. However, if $s \in S$, then for every $x \in L$, we have

$$xs + s = xs \cdot s + s,$$

that is, $xs\theta s$. Thus, S is the zero of the quotient lattice L/θ and the proof is complete.

The converse of the theorem is, unfortunately, not true: the non-standard ideal $a^\triangledown$ of the lattice depicted in Figure 4 is the kernel ideal of its homomorphism onto a two-element chain. However, some comfort is provided by the next theorem.

THEOREM 10. (Grätzer–Schmidt). *The kernel ideal of any congruence on a relatively complemented lattice is a standard ideal.*

Proof. Let L be a relatively complemented lattice and S the kernel ideal of its homomorphism φ. If $a, b \in L$, $s \in S$ and $a \leqslant b + s$, then we denote by z a complement of the element ab relative to $[abs, a]$. Then

$a = ab + z$ and

$$\varphi(z) = \varphi(z(b+s)) = \varphi(z)\,\varphi(b+s) = \varphi(z)\,\varphi(b)$$

$$= \varphi(za)\,\varphi(b) = \varphi(z(ab)) = \varphi(abs) = 0,$$

that is, $z \in S$.

A filter S of the lattice L is called *standard* if for any $a, b \in L$ and $s \in S$, the inequality $a \geqslant bs$ implies $a = xt$, where $x \geqslant b$ and $t \in S$. By the duality principle, Theorem 9 implies the next theorem.

THEOREM 11. *Every standard filter S of the lattice L is the kernel filter of the congruence θ, defined by the condition*

$$x\theta y \underset{\text{def}}{=} \boxed{xy = (x+y)\,s \text{ for some } s \in S.}$$

In connection with the fact that not every lattice is complete, the criteria for completeness are of interest, two of which are set forth below.

THEOREM 12. *If every sublattice of a lattice L with zero has a least upper bound, then L is complete.*

Proof. It is immediate from the hypothesis that L contains a unit element. If A is a non-void subset of L, then in view of the existence of zero, the lower cone $A^{\triangledown}$ is not void. It is easily verified that $A^{\triangledown}$ is a sublattice. By Theorem 1.8 we have

$$\inf A = \sup A^{\triangledown},$$

and then it remains to apply only Theorem 3.1.

THEOREM 13. (Davis). *If every isotone mapping of the lattice L into itself has a fixed point, then it is complete.*

Proof. As a preliminary, we establish

LEMMA 1. *If C is a chain in L, then there exist* $\sup_L C$ *and* $\inf_L C$.

For the proof, we make use of Theorem 2.3 and embed C into a maximal chain D. By Theorem 3.3, there is an element $c = \sup_D C$. If $v \in L$ and $v \geqslant x$ for all $x \in C$, then $cv \geqslant x$ for all $x \in C$. If $z \in D$ and $z \geqslant c$, then $z \geqslant cv$. If $z \in D$ and $z < c$, then $z \leqslant x$ for some $x \in C$; therefore $cv \geqslant x \geqslant z$. Thus, the element cv is comparable with all elements of D and in view of D being a maximal chain, belongs to it. Hence, $v \geqslant cv \geqslant c$, that is, $c = \sup_L C$.

The second statement follows by duality.

LEMMA 2. *L contains zero and a unit element.*

For proof we make use of Theorem 2.3 and select in L some maximal chain C. By Lemma 1, there exists $w = \sup C$. If $x \in L$ and $x \nleqslant w$, then $x + w > w$. It follows that $x + w \overline{\in} C$ and by adjoining to C the element $x + w$, we again get a chain. This, however, contradicts the maximality of C. Consequently, $w = 1$. The existence of zero is proved by duality.

We now proceed to prove the theorem and consider a non-void subset A of the lattice L. By Lemma 2, the lower cone $A^{\triangledown}$ is not void. We draw upon Theorem 2.3 and select in $A^{\triangledown}$ a maximal chain C. In view of Lemma 1, there exists $a = \sup C$. Since $A \subseteq C^{\vartriangle}$, it follows that $A \subseteq a^{\vartriangle}$, by the definition of a LUB. Hence $a \in A^{\triangledown}$. If $x \in A^{\triangledown}$, then $a + x \in A^{\triangledown}$ and $a + x \geqslant a$. If $a + x > a$, then the chain C admits extension, violating its maximality. However, if $a + x = a$, then $x \leqslant a$. Thus, $a = \inf A$ and it only remains to apply Theorem 3.1.

In connection with a complete lattice this lemma can be used for a definition. A sublattice H of L is called *complete*, if for every subset $A \subseteq H$, the existence of $a = \sup_L A$ $[a = \inf_L A]$ implies that $a \in H$. As already noted, in this situation we have $a = \sup_H A$ $[a = \inf_H A]$. An interval can be taken as an example of a complete sublattice. The intersection of any sets of a complete sublattice is evidently a complete sublattice. Hence, among the complete sublattices containing a given set $\mathfrak{M}$, there is a least sublattice of which we say that it is *completely generated by* $\mathfrak{M}$. For example, the unit interval [0, 1] of real numbers is completely generated by the open interval (0, 1). An ideal (filter) is called *complete*, if it is a complete sublattice.

A homomorphism φ of L into a lattice L' is called *complete* if for every subset $A \subseteq L$, for which there exists $\sup_L A$ $[\inf_L A]$, we have $\varphi(\sup_L A) = \sup_{L'} \varphi(A)$ $[\varphi(\inf_L A) = \inf_{L'} \varphi(A)]$. From Theorem 1.5, it is easy to deduce that an isomorphism is a complete homomorphism. As an example of an incomplete homomorphism we can take the mapping of real interval [0, 1] into a two-element chain, defined by the conditions

$\varphi(1) = 1$ and $\varphi(x) = 0$ if $0 \leqslant x < 1$. It is easy to verify that the kernel ideal [filter] of a complete homomorphism is complete.

THEOREM 14. *If φ is a complete homomorphism of a complete lattice L onto a lattice L' and if $a' \in L'$, then the set $I = \{x \mid x \in L, \varphi(x) = a'\}$ is an interval.*

Proof. Let $a = \inf I$ and $b = \sup I$. If $x \in [a, b]$, then $a' = \varphi(a) \leqslant \varphi(x) \leqslant \varphi(b) = a'$, that is, $[a, b] \subseteq I$. The reverse inclusion is obvious.

Note that for this conclusion it is essential that the homomorphism be complete. In the homomorphism adduced above of the interval [0, 1] onto a two-element chain, the complete pre-image of zero is not an interval.

Exercises

1. Show that the following properties of elements a and b of an arbitrary lattice are equivalent:

$$(a)\ a \leqslant b; \qquad (b)\ a \leqslant ab; \qquad (c)\ a + b \leqslant b.$$

2. Any minimal (maximal) element of a lattice is zero (unity).
3. If $a + b + c = abc$, then $a = b = c$.
4. In every lattice, the equality

$$(ab + ac)(ab + bc) = ab$$

holds.

5. Show that the following properties of a lattice L are equivalent: (*a*) L is a chain; (*b*) all subsets in L are sublattices; (*c*) every homomorphism of a poset $\bar{L}$ is a lattice homomorphism; (*d*) $a = bc$ implies $a = b$ or $a = c$.
6. A subset I of the lattice L is an ideal iff the inclusions $a, b \in I$ and $a + b \in I$ hold simultaneously.
7. Ideals form the precise class of subsets occurring as a complete pre-image of zero for some $\cup$-homomorphism (a mapping φ of L into the lattice L' is called a $\cup$-*homomorphism* if $\varphi(a + b) = \varphi(a) + \varphi(b)$).
8. Every prime ideal serves as the kernel of some lattice homomorphism (an ideal P is called *prime*, if $ab \in P$ implies $a \in P$ or $b \in P$).
9. A sublattice P of L is a prime ideal iff $L \backslash P$ is a prime filter (a filter F is called *prime* if $a + b \in F$ implies $a \in F$ or $b \in F$).
10. Show that the ideals of any arbitrary lattice L form a lattice and that the completeness of this lattice is equivalent to the presence of zero in L.
11. If a lattice satisfies the maximum condition, then all its ideals are principal. Formulate and prove the dual statement.

12. Any lattice, to which we add not more than three elements, can be embedded to a lattice with 0 and 1, in which every element has a complement.

13. A lattice with 1, satisfying the minimum condition, is complete.

14. For the completeness of a lattice, it is necessary and sufficient that all ideals satisfying the condition $I^{\Delta\nabla} = I$, are principal.

15. An ordered product of lattices is a lattice.

16. The ordered sum of the lattices P_α, $\alpha \in L$, is a lattice iff L is a lattice and $\alpha + \beta \neq \alpha, \beta$ implies the existence of zero in the lattice $P_{\alpha+\beta}$ and unit in the lattice $P_{\alpha\beta}$.

5

FREE LATTICES

In contrast to an arbitrary algebra, the notion of a free lattice permits a much wider scope with regard to the introduction of an order in the underlying set and examining various aspects of the preservation of properties obtained for a poset when embedding it in the free lattice.

Let P be a poset. A subset A of P is called *summable* [*multiplicative*], if there exists sup A [inf A]. In P select some systems $\mathfrak{A}$ and $\mathfrak{B}$ of summable and multiplicative sets, respectively. A lattice F is called the $(\mathfrak{A}, \mathfrak{B})$-*free extension* of P if

(i) there exists an isomorphism θ of P into F;

(ii) F is generated by the set $\theta(P)$;

(iii) if $M \in \mathfrak{A}$, then $\sup_F(\theta(M))$ exists and $\sup_F(\theta(M)) = \theta(\sup_P M)$;

(iv) if $N \in \mathfrak{B}$, then $\inf_F(\theta(N))$ exists and $\inf_F(\theta(N)) = \theta(\inf_P N)$;

(v) if φ is an isotone mapping of P into some lattice L; moreover, if for any $M \in \mathfrak{A}$ and $N \in \mathfrak{B}$ there exists $\sup_L(\varphi(M))$ and $\inf_L(\varphi(N))$ coinciding with $\varphi(\sup_P M)$ and $\varphi(\inf_P N)$, respectively, then there is a homomorphism ψ of F into L, such that $\psi(\theta(x)) = \varphi(x)$ for all $x \in P$.

The *free* and *complete free extensions* are important in particular cases. The former arises when $\mathfrak{A}$ consists of all summable and $\mathfrak{B}$ of all multiplicative pairs of the set P; the latter occurs when both $\mathfrak{A}$ and $\mathfrak{B}$ coincide with the set of all comparable pairs. The free and complete free extensions of a trivial poset coincide. The corresponding lattice is called the *free lattice.* The free extension of a cardinal sum of lattices is called their *free product.* All these definitions are in complete agreement with usage in the theory of universal algebra.

For solving the problem of the existence of the $(\mathfrak{A}, \mathfrak{B})$-free extension of a poset P, we introduce the notion of a *word,* namely, the words of weight zero are designated as elements of P. The words of weight n are defined as the expressions $A + B$ and AB, where A and B are the words of smaller weight. The words A and B and all their subwords are called the *subwords* of $A + B$ and AB. Of course, the weight of a word in this connection is not defined uniquely. However, among the weights corresponding to a given word A, there exists a least, denoted by the symbol $w(A)$. In the following, we shall call this the *weight of the word A.* The words A and B of weight 0 are regarded equal (in symbols $A \equiv B$), if they are equal as the elements of P. The equalities $A + B \equiv C + D$ and $AB \equiv CD$, by definition, hold iff $A \equiv C$ and $B \equiv D$. The words of weight 0 and of the form AB are called *$\cup$-irreducible,* while the words of weight 0 and of the form $A + B$ are called *$\cap$-irreducible.* The representation of A in the form

$$A \equiv A_1 + A_2 + \cdots + A_m, \tag{5.1}$$

where the A_i are $\cup$-irreducible words, and having some distribution of parentheses, is called a *$\cup$-canonical representation.* The representation of A in the form

$$A \equiv A_1 A_2 \ldots A_m, \tag{5.2}$$

where the A_i are $\cap$-irreducible words, and having some distribution of parentheses, is called a *$\cap$-canonical representation.* It is easy to see that for every word, both these representations exist and are defined uniquely. However, at least for one of them $m = 1$. If $m > 1$, then the weight of A_i is less than that of A. The *value* $v(A)$ of a word A of weight 0 is the element A. The *value* $v(A)$ of A with $\cup$-canonical (or $\cap$-canonical) representation (5.1) (or (5.2)) is defined when for all i, the values $v(A_i)$ are defined and the set $\{v(A_i) \mid i = 1, \ldots, m\}$ belongs to the system $\mathfrak{A}$ (or $\mathfrak{B}$). Subject to these conditions, we put

$$v(A) = \sup_P \{v(A_i)\}$$

(or $v(A) = \inf_P \{v(A_i)\}$).

So, the value of a word is a uniquely defined element in P. However, not every word has a value.

On the set of words we define the relations $\underset{n}{\leqslant}$ ($n = 1, 2, \ldots$) by putting

$$A \underset{1}{\leqslant} B \underset{\text{def}}{=} \boxed{A \equiv B \text{ or } v(A) \leqslant_P v(B),}$$

and if $n > 1$

$A \underset{n}{\leqslant} B \underset{\text{def}}{=}$

(i) $A \underset{n-1}{\leqslant} C$ and $C \underset{n-1}{\leqslant} B$ for some word C; or

(ii) $A \equiv A_1 + \cdots + A_m$ is a $\cup$-canonical representation and $A_i \underset{n-1}{\leqslant} B$ for all $i = 1, \ldots, m$; or

(iii) $A \equiv A_1 \ldots A_m$ is a $\cap$-canonical representation and $A_i \underset{n-1}{\leqslant} B$ for some $i = 1, \ldots, m$; or

(iv) $B \equiv B_1 \ldots B_m$ is a $\cap$-canonical representation and $A \underset{n-1}{\leqslant} B_i$ for all $i = 1, \ldots, m$; or

(v) $B \equiv B_1 + \cdots + B_m$ is a $\cup$-canonical representation and $A \underset{n-1}{\leqslant} B_i$ for some $i = 1, \ldots, m$.

PROPOSITION 1. *$A \underset{n}{\leqslant} A$ for all n.*

Indeed for $n = 1$, the lemma holds by definition. Now if $A \underset{n-1}{\leqslant} A$ and $A \underset{n-1}{\leqslant} A$, then by applying rule (i), we have $A \underset{n}{\leqslant} A$.

PROPOSITION 2. *If $A \underset{n}{\leqslant} B$ and $m \geqslant n$, then $A \underset{m}{\leqslant} B$.*

Indeed from the hypotheses and Lemma 1, it follows that $A \underset{n}{\leqslant} B$ and $B \underset{n}{\leqslant} B$. Applying rule (i) we get $A \underset{n+1}{\leqslant} B$. The rest follows by induction.

Now, put

$$A \leqslant B \underset{\text{def}}{=} \boxed{A \underset{n}{\leqslant} B \text{ for some } n.}$$

PROPOSITION 3. *The relation $\leqslant$ is reflexive and transitive.*

Indeed the reflextivity of $\leqslant$ is immediate from Proposition 1. If $A \leqslant B$ and $B \leqslant C$, then in view of Proposition 2, we can assume that $A \underset{n}{\leqslant} B$ and $B \underset{n}{\leqslant} C$ for some n. Applying rule (i), we get $A \underset{n+1}{\leqslant} B$, that is, $A \leqslant B$.

PROPOSITION 4. $AB \leqslant A \leqslant A + B$ *and* $AB \leqslant B \leqslant A + B$.

In fact, if $C \equiv A + B$, and

$$A \equiv A_1 + \cdots + A_m,$$

and

$$B \equiv B_1 + \cdots + B_n$$

are $\cup$-canonical representations, then, by Proposition 1 and rule (v), $C \geqslant A_i$ for all $i = 1, \ldots, m$. In view of rule (ii), it thereby follows that $C \geqslant A$. Similarly, it is verified that $C \geqslant B$. The inequalities $AB \leqslant A$, B are proved by duality.

PROPOSITION 5. *If* $C \leqslant A, B$, *then* $C \leqslant AB$.

For the proof, consider the $\cap$-canonical representations

$$A \equiv A_1 \ldots A_m,$$

and

$$B \equiv B_1 \ldots B_n.$$

Then

$$AB \equiv (A_1 \ldots A_m)(B_1 \ldots B_n)$$

is also a $\cap$-canonical representation. From Propositions 3 and 4, it follows that $C \leqslant A_i, B_j$ ($i = 1, \ldots, m$; $j = 1, \ldots, n$), whence, in view of rule (iv), we have $C \leqslant AB$.

Dually, the next proposition is proved.

PROPOSITION 6. *If* $A, B \leqslant C$, *then* $A + B \leqslant C$.

Proposition 3 and Theorem 1.3 allow us to consider a quotient set F of the set of words with respect to an equivalence relation ρ, defined by the conditions

$$A \rho B \underset{\text{def}}{=} \boxed{A \leqslant B \text{ and } B \leqslant A.}$$

By Theorem 1.3, the relation $\leqslant$ induces an order on F. In the following, the word A itself and its image under the natural mapping onto will be denoted by one and the same symbol. The expression $A = B$ denotes the equality of elements in F. For denoting the equality of words A and B, as earlier we shall employ the expression $A \equiv B$.

From Propositions 4–6 the next proposition is immediate.

Proposition 7. *The poset F is a lattice, and $A + B = \sup \{A, B\}$ and $AB = \inf \{A, B\}$.*

Furthermore, we consider a poset Q of integer-valued rows indicated in Theorem 2.10 and we pass on to the next proposition.

Proposition 8. *If*

$$q = (m_1, \ldots, m_n) \in Q,$$

$$A_0 \underset{m_1}{\leqslant} A_1 \underset{m_2}{\leqslant} \ldots \underset{m_{n-1}}{\leqslant} A_{n-1} \underset{m_n}{\leqslant} A_n,$$

and if the values $v(A_1), \ldots, v(A_{n-1})$ do not exist, then the row

$$r = (r_1, \ldots, r_s) < q,$$

such that

$$A_0 \underset{r_1}{\leqslant} D_1 \underset{r_2}{\leqslant} \cdots \underset{r_{s-1}}{\leqslant} D_{s-1} \underset{r_s}{\leqslant} A_n,$$

exists in the following cases: (1) $A_k \equiv A_{k+1}$ *for some* k; (2) *for some* k *the inequality* $\underset{m_k}{\leqslant}$ *is established with the help of rule* (i); (3) *for some* $k \neq n$ *the inequality* $\underset{m_k}{\leqslant}$ *is established with the aid of the rules* (iv) *and* (v), *these rules being not applicable for establishing the inequality* $\underset{m_j}{\leqslant}$ *for* $j > k$; (4) *for some* $k \neq 1$, *the inequality* $\underset{m_k}{\leqslant}$ *is established with the aid of the rules* (ii) *and* (iii), *these rule being not applicable for establishing the inequalities* $\underset{m_j}{\leqslant}$ *for* $j < k$.

Indeed in case (1) we have

$$A_0 \underset{m_1}{\leqslant} \dots \underset{m_{k-1}}{\leqslant} A_{k-1} \underset{m_{k+1}}{\leqslant} A_{k+1} \underset{m_{k+2}}{\leqslant} \dots \underset{m_n}{\leqslant} A_n,$$

which allows us to write

$$r = (m_1, \dots, m_{k-1}, \ m_{k+1}, \dots, m_n).$$

In case (2) one must have

$$A_{k-1} \underset{m_{k-1}}{\leqslant} C \underset{m_{k-1}}{\leqslant} A_k$$

for some word C. Consequently,

$$A_0 \underset{m_1}{\leqslant} \dots \underset{m_{k-1}}{\leqslant} A_{k-1} \underset{m_{k-1}}{\leqslant} C \underset{m_{k-1}}{\leqslant} A_k \underset{m_{k+1}}{\leqslant} \dots \underset{m_n}{\leqslant} A_n,$$

and we can write

$$r = (m_1, \dots, m_{k-1}, m_k - 1, m_k - 1, m_{k+1}, \dots, m_n).$$

In case (3) by applying rule (iv) we have the $\cap$-canonical representation

$$A_k \equiv B_1 \dots B_p.$$

Of course, we can assume that $p \neq 1$ and that cases (1) and (2) do not hold. Hence, the inequality $\underset{m_{k+1}}{\leqslant}$ can be established with the aid of rules (ii), (iii), (iv) or (v). However, the application of rule (iv) or (v) contradicts the choice of the index k. On applying rule (ii) we have $A_k \underset{m_{k+1}-1}{\leqslant} A_{k+1}$ which allows us to set

$$r = (m_1, \dots, m_k, m_{k+1} - 1, m_{k+2}, \dots, m_n).$$

Thus, it remains to be shown that the inequality $\underset{m_{k+1}}{\leqslant}$ is established with the aid of rule (iii). But, then, for some i we have $B_i \underset{m_{k+1}-1}{\leqslant} A_{k+1}$, and for all i we have $A_{k-1} \underset{m_k-1}{\leqslant} B_i$. Consequently, for a suitable index i we have

$$A_0 \underset{m_1}{\leqslant} \dots \underset{m_k-1}{\leqslant} A_{k-1} \underset{m_k-1}{\leqslant} B_i \underset{m_{k+1}-1}{\leqslant} A_{k+1} \underset{m_{k+2}}{\leqslant} \dots \underset{m_n}{\leqslant} A_n,$$

enabling us to write

$$r = (m_1, \ldots, m_{k-1}, m_k - 1, m_{k+1} - 1, m_{k+2}, \ldots, m_n).$$

However, if the inequality $\underset{m_k}{\leqslant}$ is established with the help of rule (v), then we have the $\cup$-canonical representation

$$A_k \equiv B_1 + \cdots + B_p.$$

This shows that we may confine ourselves to the case when the inequality $\underset{m_{k+1}}{\leqslant}$ is established by applying rule (ii). Consequently, $A_{k-1} \underset{m_k - 1}{\leqslant} B_i$ for some i and $B_i \underset{m_{k+1}-1}{\leqslant} A_{k+1}$ for all i. Hence, for a suitable i, we have

$$A_0 \underset{m_1}{\leqslant} \ldots \underset{m_{k-1}}{\leqslant} A_{k-1} \underset{m_k - 1}{\leqslant} B_i \underset{m_{k+1}-1}{\leqslant} A_{k+1} \underset{m_{k+2}}{\leqslant} \ldots \underset{m_n}{\leqslant} A_n,$$

and the row

$$r = (m_1, \ldots, m_{k-1}, m_k - 1, m_{k+1} - 1, m_{k+2}, \ldots, m_n).$$

Case (4) is considered dually.

PROPOSITION 9. *If*

$$(m_1, \ldots, m_n) \in Q,$$

$$A_0 \underset{m_1}{\leqslant} A_1 \underset{m_2}{\leqslant} \ldots \underset{m_n}{\leqslant} A_n,$$

and if the values $v(A_0)$ *and* $v(A_n)$ *exist, then*

$$v(A_0) \leqslant_P v(A_n).$$

For the proof, consider the set T of all those rows, for which Proposition 9 does not hold. If T is void, there is nothing to prove. In the opposite case, by Theorem 2.10, T contains a minimal element $q = (m_1, \ldots, m_n)$. So

$$v(A_0) \nleqslant_P v(A_n). \tag{5.3}$$

If $q = 1$, then, by definition of the relation $\underset{1}{\leqslant}$, we have $A_0 \equiv A_1$ or

$v(A_0) \leqslant_P v(A_1)$, which is inconsistent with the relation (5.3). If there exists a value $v(A_i)$ for $i \neq 0, n$, then, in view of the minimality of q and the inequalities $(m_1, \ldots, m_i) < q$ and $(m_{i+1}, \ldots, m_n) < q$ being obvious, we have

$$v(A_0) \leqslant_P v(A_i) \leqslant_P v(A_n),$$

again contradicting the relation (5.3). Thus, it is found that the hypotheses of Proposition 8 are satisfied and the minimality of q implies the inequalities $m_1, m_n > 1$ and the presence of one of the following cases:

(i) (*a*) the inequality $\underset{m_1}{\leqslant}$ is established with the help of rule (ii);

(ii) (*b*) the inequality $\underset{m_1}{\leqslant}$ is established with the help of rule (iii);

(iii) (*c*) the inequality $\underset{m_n}{\leqslant}$ is established with the help of rule (iv);

(iii) (*d*) the inequality $\underset{m_n}{\leqslant}$ is established with the aid of rule (v);

Case (i) (*a*). We have the $\cup$-canonical representation

$$A_0 \equiv B_1 + \cdots + B_m,$$

where $B_i \underset{m_1-1}{\leqslant} A_1$ for all i. Since

$$v(A_0) = v(B_1) + \cdots + v(B_m),$$

hence, for all i, there exists $v(B_i)$. Since

$$(m_1 - 1, m_2, \ldots, m_n) < q,$$

therefore, $v(B_i) \leqslant_P v(A_n)$ for all i and, consequently, $v(A_0) \leqslant_P v(A_n)$, violating the relation (5.3).

Case (ii) (*b*). We have the $\cap$-canonical representation

$$A_0 \equiv B_1 \ldots B_m,$$

and, say $B_1 \underset{m-1}{\leqslant} A_1$. Since

$$v(A_0) = v(B_1) \ldots v(B_m),$$

hence, $A_0 \underset{1}{\leqslant} B_1$. Since

$$(1, m_1 - 1, m_1, \ldots, m_n) < q,$$

therefore, $v\,(A_0) \leqslant_P v\,(A_n)$, violating (5.3).

Case (iii) (*c*). We have the $\cap$-canonical representation

$$A_n \equiv B_1 \ldots B_m,$$

and $A_{n-1} \underset{m_n - 1}{\leqslant} B_i$ for all i. In addition, for all i, there exists $v\,(B_i)$. Since

$$(m_1, \ldots, m_{n-1}, m_n - 1) < q,$$

hence, $v\,(A_0) \leqslant_P v\,(B_i)$ for all i, whence $v\,(A_0) \leqslant_P v\,(A_n)$, violating (5.3).

Case (iii) (*d*). We have the $\cup$-canonical representation

$$A_n \equiv B_1 + \cdots + B_m$$

and, say $A_{n-1} \underset{m_n - 1}{\leqslant} B_1$. In addition,

$$v\,(A_n) = v\,(B_1) + \cdots + v\,(B_m),$$

whence $B_1 \underset{1}{\leqslant} A_n$. Since

$$(m_1, \ldots, m_{n-1}, m_n - 1, 1) < q,$$

we have $v\,(A_0) \leqslant_P v\,(A_n)$, contradicting (5.3).

PROPOSITION 10. *If*

$$q = (m_1, \ldots, m_n) \in Q,$$

$$A_0 \underset{m_1}{\leqslant} A_1 \underset{m_2}{\leqslant} \ldots \underset{m_n}{\leqslant} A_n,$$

if A_n *has the* $\cup$-*canonical representation*

$$A_n \equiv D_1 + \cdots + D_t$$

and there exists a value $v\,(A_0)$, *then either there exists a value* $v\,(A_n)$ *or* $A_0 \leqslant_F D_j$ *for all* $j = 1, \ldots, t$.

For proof, as we did in the case of Proposition 9, we can assume that q is minimal among the elements of a poset Q, for which the proposition in hand does not hold. If $q = (1)$, then there is a value $v\,(A_n)$. If there is a value $v\,(A_i)$ for some $i \neq 0, n$, then, since $(m_{i+1}, \ldots, m_n) < q$, for some j we have $A_0 \leqslant_F A_i \leqslant_F D_j$. If the inequality $\underset{m_n}{\leqslant}$ is established by rule (iv), then $A_{n-1} \underset{m_n-1}{\leqslant} A_n$, a contradiction to the choice of q. In view of Proposition 8, it remains to show that either the inequality $\underset{m_n}{\leqslant}$ is established by rule (v), or the inequality $\underset{m_1}{\leqslant}$ is established by means of the rules (ii) or (iii). From the former proposition, we have $A_{n-1} \underset{m_n-1}{\leqslant} D_j$ for some j and, consequently, $A_0 \leqslant_F D_j$. The latter proposition implies the presence of the representation

$$A_0 \equiv C_1 + \cdots + C_s,$$

or

$$A_0 \equiv C_1 \ldots C_s,$$

where $C_i \underset{m_1-1}{\leqslant} A_1$ for all i or for some i, respectively. In view of the existence of the value $v\,(C_i)$ for this i and a suitable i, we have the inequalities $C_i \leqslant_F D_j$, implying that $A_0 \leqslant_F D_j$.

The next proposition is proved by duality.

PROPOSITION 11. *If* $A_0 \leqslant_F A_n$, *if* A_0 *has the* $\cap$*-canonical representation* $A_0 \equiv C_1 \ldots C_s$ *and there exists a value* $v\,(A_n)$, *then either there exists a value* $v\,(A_0)$ *or* $C_i \leqslant_F A_n$ *for some* $i = 1, \ldots, s$.

Further, we have

PROPOSITION 12. *If*

$$q = (m_1, \ldots, m_n) \in Q,$$

$$A_0 \equiv C_1 \ldots C_s$$

and

$$A_n \equiv D_1 + \ldots + D_t$$

are the $\cap$*- and* $\cup$*-canonical representations, if the values* $v\,(A_0)$ *and* $v\,(A_n)$ *do not exist and if*

$$A_0 \underset{m_1}{\leqslant} A_1 \underset{m_2}{\leqslant} \dots \underset{m_n}{\leqslant} A_n,$$

then $C_i \leqslant D_j$ *for some* i *and* j.

For proof, we assume as earlier that q is minimal among the elements of a poset Q, for which the proposition in hand does not hold. The assumption that the inequality $\underset{m_1}{\leqslant}$ is established by rule (ii) or $\underset{m_n}{\leqslant}$ is established by rule (iv), leads to a contradiction with the choice of q. If $v\,(A_k)$ exists for some k, then, in view of Propositions 10 and 11, for some i and we have

$$C_i \leqslant A_k \leqslant D_j.$$

In the contrary case, from Proposition 8 it follows that either the inequality $\underset{m_1}{\leqslant}$ is established by rule (iii), or $\underset{m_n}{\leqslant}$ is established by rule (v). In the former case for some i, $C_i \underset{m_1-1}{\leqslant} A_1$ holds. If $v\,(C_i)$ exists, then $C_i \leqslant D_j$ for some j, in view of Proposition 10. However, if $v\,(C_i)$ does not exist, then we get the same inequality, noting that C_i is itself the free $\cap$-irreducible representation and that

$$(m_1 - 1,\ \ m_2, \dots, m_n) < q.$$

The latter case is disposed of by duality.

PROPOSITION 13. *If* $A = \sup_F \{A_1, \dots, A_n\}$ *and the values* $v\,(A)$, $v\,(A_1), \dots, v\,(A_m)$ *exist, then*

$$v\,(A) = \sup_P \{v\,(A_1), \dots, v\,(A_m)\}.$$

Indeed, since $A \geqslant_F A_i$ for all i, hence, by Proposition 9 $v(A) \geqslant_P v(A_i)$ for all i. If $x \in P$ and $x \geqslant_P v\,(A_i)$ for all i, then, regarding x as a word of weight zero, we have $x \underset{1}{\geqslant} A_i$ for all i. Consequently, $x \geqslant_F A$ and a repeated application of Proposition 9 yields $x \geqslant_P v\,(A)$. Thus,

$$v\,(A) = \sup_P \{v\,(A_1), \dots, v\,(A_m)\}.$$

The next proposition is proved by duality.

Proposition 14. *If* $A = \inf_F \{A_1, \ldots, A_m\}$ *and the values* $v(A)$, $v(A_1), \ldots, v(A_m)$ *exist, then*

$$v(A) = \inf_P \{v(A_1), \ldots, v(A_m)\}.$$

Theorem 1. (Dilworth–Dean). *If* $\mathfrak{A}$ *and* $\mathfrak{B}$ *consist only of finite subsets, then the lattice F is the* $(\mathfrak{A}, \mathfrak{B})$*-free extension of a poset P.*

Proof. We shall successively verify that the lattice F possesses the properties (i)–(v) entering in the definition of $(\mathfrak{A}, \mathfrak{B})$-free extension.

(i) If $a \in P$, then we denote by $\theta(a)$ the word a of weight zero. If a, $b \in P$ and $a \leqslant_P b$, then $v(\theta(a)) = a$ and $v(\theta(b)) = b$, whence, $\theta(a) \underset{1}{\leqslant} \theta(b)$. However, if $\theta(a) \leqslant \theta(b)$, then

$$\theta(a) \underset{m_1}{\leqslant} A_1 \underset{m_2}{\leqslant} \ldots \underset{m_n}{\leqslant} \theta(b),$$

whence, in view of Proposition 9, it follows that

$$a = v(\theta(a)) \leqslant_P v(\theta(b)) = b.$$

In particular, if $\theta(a) = \theta(b)$, then $a \leqslant_P b \leqslant_P a$, whence $a = b$. So θ is the isomorphism of P into F.

(ii) This follows trivially from Proposition 7.

(iii) If $M \in \mathfrak{A}$, then M consists of a finite number of elements. Let these elements be $a_1, \ldots, a_m$. We compose the word

$$A \equiv \theta(a_1) + \cdots + \theta(a_m)$$

(existing for some distribution of parentheses). Since $v(A) = v(\theta(v(A)))$, then $A \underset{1}{\leqslant} \theta(v(A)) \underset{1}{\leqslant} A$, that is, $A = \theta(v(A))$. Therefore, by means of Propositions 7 and 13, we deduce

$$\sup_F \theta(M) = A = \theta(v(A)) = \theta(\sup_P M).$$

(iv) This is verified by duality.

(v) Let φ and L have the same meaning as in the formulation of this property. For the word A of weight zero, put $\psi(A) = \varphi(v(a))$. For the word A of weight $w > 0$ consider its $\cup$- and $\cap$-canonical representations

$$A \equiv A_1 + \cdots + A_h$$

and

$$A \equiv A_1' \ldots A_{h'}'.$$

In precisely one of these representations, there appears on the right-hand side of the equality a word having weight less than w. If this holds for the former representation, then we write

$$\psi(A) = \sup_L \{\psi(A_i) \mid i = 1, \ldots, h\},$$

but if for the latter, then

$$\psi(A) = \inf_L \{\psi(A_i') \mid i = 1, \ldots, h'\}.$$

We shall see that if $v(A)$ exists, then

$$\psi(A) = \varphi(v(A)). \tag{5.4}$$

In fact for the words of weight 0, this holds by definition. For the words of a larger weight, using the induction hypothesis and observing that in the considered case the collection of values of the terms of a $\cup$-canonical (correspondingly $\cap$-canonical) representation lies in $\mathfrak{A}$ (correspondingly $\mathfrak{B}$), we have

$$\psi(A) = \sup_L \{\psi(A_i)\} = \sup_L \{\varphi(v(A_1))\}$$

$$= \varphi(\sup_P \{v(A_i)\}) = \varphi(v(A)),$$

or

$$\psi(A) = \inf_L \{\psi(A_i')\} = \inf_L \{\varphi(v(A_i')\} = \varphi(\inf_P \{v(A_i')\} = \varphi(v(A)).$$

Furthermore, we shall establish that $A \underset{n}{\leqslant} B$ implies $\psi(A) \leqslant_L \psi(B)$. In fact, if $n = 1$, then for $A \equiv B$, we evidently have $\psi(A) \leqslant_L \psi(B)$. However, if $v(A) \leqslant_P v(B)$, then by means of (5.4) we obtain

$$\psi(A) = \varphi(v(A)) \leqslant \varphi(v(B)) = \psi(B).$$

If $A \underset{n}{\leqslant} B$, then in case of an application of rule (i), by using induction, we get

$$\psi(A) \leqslant_L \psi(C) \leqslant_L \psi(B).$$

If rule (ii) or (iv) is applied for all $i = 1, \ldots, m$, we have $\psi(A_i) \leqslant_L \psi(B)$ or $\psi(A) \leqslant_L \psi(B_i)$ respectively. Hence,

$$\psi(A) = \sup_L \{\psi(A_i)\} \leqslant_L \psi(B),$$

or

$$\psi(A) \leqslant_L \inf_L \{\psi(B_i)\} = \psi(B)$$

holds. If, however, rule (iii) or (v) is applied, then, for some numbers i_0 and j_0, we have $\psi(A_{i_0}) \leqslant_L \psi(B)$ or $\psi(A) \leqslant_L \psi(B_{j_0})$, respectively. Therefore,

$$\psi(A) = \inf_L \{\psi(A_i)\} \leqslant_L \psi(A_{i_0}) \leqslant_L \psi(B),$$

or

$$\psi(A) \leqslant_L \psi(B_{j_0}) \leqslant_L \sup\{\psi(B_j)\} = \psi(B).$$

Thus, $A \leqslant_F B$ implies $\psi(A) \leqslant_L \psi(B)$. In particular, if $A = B$, then $\psi(A) = \psi(B)$. Consequently, ψ is an isotone map of the lattice F into the lattice L, and $\psi(\theta(x)) = \varphi(v(\theta(x))) = \varphi(x)$ for all $x \in P$. If $C \equiv A + B$, then we write the $\cup$-canonical representations

$$A \equiv A_1 + \cdots + A_m,$$

and

$$B \equiv B_1 + \cdots + B_n.$$

Taking note of Theorem 1.4, we ge

$$\psi(A + B) = \sup_L \{\psi(A_1), \ldots, \psi(A_m), \psi(B_1), \ldots, \psi(B_n)\}$$

$$= \sup_L \{\sup_L \{\psi(A_i)\}, \sup_L \{\psi(B_j)\}\}$$

$$= \sup_L \{\psi(A), \psi(B)\} = \psi(A) + \psi(B).$$

Dually, we arrive at

$$\psi(AB) = \psi(A)\,\psi(B).$$

Consequently, ψ is the lattice homomorphism. We denote by $FL(n)$ the free extension of a trivial poset, consisting of n elements $x_1, \ldots, x_n$. The elements $x_1, \ldots, x_n$ are called *free generators*. It is easily verified that $FL(2)$ is the four-element lattice of Figure 5. The elements a and b are its free generators.

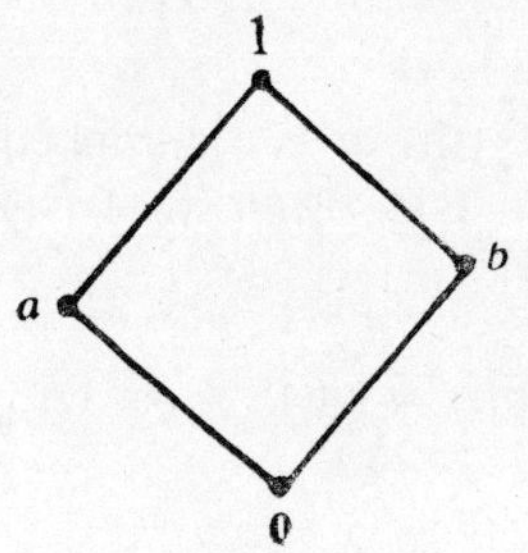

Figure 5.

THEOREM 2. (Whitman). *The elements $u_1, \ldots, u_m$ of a lattice $F = FL(n)$ are free generators of the sublattice they generate iff for every $i = 1, \ldots, m$ we have*

$$u_i \nleq u_1 + \cdots + u_{i-1} + u_{i+1} + \cdots + u_m, \tag{5.5}$$

and

$$u_i \ngeq u_1 \ldots u_{i-1} u_{i+1} \ldots u_m. \tag{5.6}$$

Proof. We denote by U the sublattice of F generated by the elements

$u_1, \dots, u_m$. If $U = FL(m)$ with the free generators $u_1, \dots, u_m$, but, say,

$$u_m \leqslant u_1 + \cdots + u_{m-1},$$

or

$$u_m \geqslant u_1 \dots u_{m-1},$$

then, mapping the set $u_1, \dots, u_m$ into $FL(2)$ (Figure 5), we put $\varphi(u_m) = a$ and $\varphi(u_1) = \cdots = \varphi(u_{m-1}) = b$. In view of property (v) of the free extensions, we can treat φ as the homomorphism of U into the lattice $FL(2)$. This, however, leads to the false relations

$$a = \varphi(u_m) \leqslant_{FL(2)} \varphi(u_1 + \dots + u_{m-1}) = b,$$

or

$$a = \varphi(u_m) \geqslant_{FL(2)} \varphi(u_1 \dots u_{m-1}) = b.$$

Let us now assume that $u_1, \dots, u_m$ satisfy the conditions (5.5) and (5.6). Consider the lattice $G = FL(m)$. By the construction of the free extension, described during the proof of Theorem 1, the lattices F and G are the quotient sets of the word sets $\mathfrak{F}$ and $\mathfrak{G}$ in the alphabets $\{x_1, \dots, x_n\}$ and $\{y_1, \dots, y_m\}$ respectively. We keep fixed in $\mathfrak{F}$ the words $U_1, \dots, U_m$ whose images under the natural mapping are formed by the elements $u_1, \dots, u_m$ and construct a mapping Φ of $\mathfrak{G}$ into $\mathfrak{F}$, by putting

$$\Phi(y_i) \equiv U_i \ (i = 1, \dots, m),$$

$$\Phi(A + B) \equiv \Phi(A) + \Phi(B)$$

and

$$\Phi(AB) \equiv \Phi(A)\,\Phi(B).$$

The symbols $=, \equiv, \underset{p}{\leqslant}, \leqslant$ will be used here in the same sense as in the proof of Theorem 1. We denote by $w(A)$ the weight of the word A. The image of A from $\mathfrak{F}$ or $\mathfrak{G}$ into the lattices F or G, respectively, is denoted by $\bar{A}$. We shall also draw upon the propositions that prefaced Theorem 1.

(*a*) *If* $A \in \mathfrak{G}$ *and* $\bar{A} \ngeqslant \bar{y}_i$, *then*

$$\bar{A} \leqslant \bar{y}_1 + \cdots + \bar{y}_{i-1} + \bar{y}_{i+1} + \cdots + \bar{y}_m.$$

In fact, the statement (*a*) is trivial, if $w(A) = 0$, since in this case $A \equiv y_j$, where $i \neq j$. However, if $w(A) > 0$, then $A \equiv B + C$ or $A \equiv BC$, where $w(B), w(C) < w(A)$. Since G is a lattice, in the former case, we have $\bar{B}, \bar{C} \ngeqslant \bar{y}_i$, whence, by induction, we get

$$\bar{B}, \bar{C} \leqslant \bar{y}_1 + \cdots + \bar{y}_{i-1} + \bar{y}_{i+1} + \cdots + \bar{y}_m.$$

The desired inequality is, evidently, immediate from these relations. In the latter case, from the hypothesis it follows that, say, $\bar{B} \ngeqslant \bar{y}_i$ and the induction hypothesis enables us to obtain the inequality

$$\bar{A} \leqslant \bar{B} \leqslant \bar{y}_1 + \cdots + \bar{y}_{i-1} + \bar{y}_{i+1} + \cdots + \bar{y}_m.$$

The next assertation is proved by duality.

(*b*) *If* $A \in \mathfrak{G}$ *and* $\bar{A} \nleqslant \bar{y}_i$, *then*

$$\bar{A} \geqslant \bar{y}_1 \ldots \bar{y}_{i-1} \bar{y}_{i+1} \ldots \bar{y}_m.$$

(*c*) *If* $A \underset{p}{\leqslant} B$, then $\overline{\Phi(A)} \leqslant \overline{\Phi(B)}$.

Indeed, if $p=1$, then by the triviality of a poset $\{y_1, \ldots, y_m\}$ we have $A \equiv B$ and, consequently, $\overline{\Phi(A)} = \overline{\Phi(B)}$. If $p > 1$, then, by applying rule (i) we have $A \leqslant C \leqslant B$ for some $C \in \mathfrak{G}$, whence $\overline{\Phi(A)} \leqslant \overline{\Phi(C)} \leqslant \overline{\Phi(B)}$ in view of the induction hypothesis. If rule (ii) is applied, then, writing the $\cup$-canonical representation $A \equiv A_1 + \cdots + A_m$, we get $A_i \underset{p-1}{\leqslant} B$ for all i. But $\Phi(A) \equiv \Phi(A_1) + \cdots + \Phi(A_m)$ and $\overline{\Phi(A_i)} \leqslant \overline{\Phi(B)}$ for all i, whence $\overline{\Phi(A)} \leqslant \overline{\Phi(B)}$. When applying rule (iii), we write the $\cap$-canonical representation $A \equiv A_1 \ldots A_m$, getting $A_i \underset{p-1}{\leqslant} B$ for some i and the induction hypothesis leads to the inequality

$$\overline{\Phi(A)} \leqslant \overline{\Phi(A_i)} \leqslant \overline{\Phi(B)}.$$

The cases of our applying the rule (iv) or (v) are disposed of along similar lines.

(*d*) *If* $B \in \mathfrak{G}$ *and* $u_i \leqslant \overline{\Phi(B)}$, *then* $\bar{y}_i \leqslant \bar{B}$.

Indeed, if $\bar{y}_i \nleqslant \bar{B}$, then by taking note of (*a*) and (*c*) we get

$$u_i = \overline{U_i} \leqslant \overline{\Phi(B)} \leqslant \overline{\Phi(y_1 + \cdots + y_{i-1} + y_{i+1} + \cdots + y_m)}$$

$$= u_1 + \cdots + u_{i-1} + u_{i+1} + \cdots + u_m,$$

which contradicts (5.5).

The assertion that follows is proved by duality.

(*e*) *If* $A \in \mathfrak{G}$ *and* $\overline{\Phi(A)} \leqslant u_i$, *then* $\bar{A} \leqslant \bar{y}_i$.

(*f*) *If* $\overline{\Phi(A)} \leqslant \overline{\Phi(B)}$, *then* $\bar{A} \leqslant \bar{B}$.

For proof, we introduce the poset Z, a lexicographic product of two chains of non-negative integers. It is easily verified that Z satisfies the minimum condition. Consider the set T of all rows (ξ, η) in Z, such that for some words A and B, where $w(A) = \xi$, $w(B) = \eta$, the statement (*f*) does not hold.

If T is void, then everything is proved. In the contrary case, we can assume that $(w(A), w(B))$ is a minimal element in T. From (*d*) and (*e*) it follows that $w(A), w(B) \neq 0$. Hence, for the elements A and B, there exist the $\cup$- or $\cap$-canonical representations

$$A \equiv A_1 + \cdots + A_s, \quad \text{or} \quad A \equiv A_1 \ldots A_s,$$

and

$$B \equiv B_1 + \cdots + B_t, \quad \text{or} \quad B \equiv B_1 \ldots B_t,$$

where $w(A_i) \leqslant w(A)$ and $w(B_i) < w(B)$. If for A, the $\cup$-canonical representation holds, then in view of (*c*)

$$\overline{\Phi(A_i)} \leqslant \overline{\Phi(A)} \leqslant \overline{\Phi(B)}$$

for all i. But since

$$(w(A_i), w(B)) < (w(A), w(B)),$$

hence, for all i there holds $\overline{A_i} \leqslant \overline{B}$ and, consequently, $\overline{A} \leqslant \overline{B}$. The case when for B there is a $\cap$-canonical representation is analysed on similar lines. Thus, in the following, we can assume that

$$\Phi(A) \equiv \Phi(A_1) \ldots \Phi(A_s),$$

and

$$\Phi(B) \equiv \Phi(B_1) + \cdots + \Phi(B_t).$$

These representations can be extended up to $\cap$- and $\cup$-canonical representations whose elements we denote by C_i and D_j, respectively. If there exists a value $v(\Phi(A))$, then $\Phi(A) \equiv x_l$ for some index l. This is possible only when $A \equiv y_h$ for some index h, since by definition of the mapping Φ, the images of all the remaining words have a nonzero weight. But, for this case the inequality $\overline{A} \leqslant \overline{B}$ holds by (*d*). If the value $v(\Phi(A))$ does not exist, then by applying Proposition 11 or 12, for suitable indices i, j and k we get

$$\Phi(A_k) \leqslant C_i \leqslant \Phi(B),$$

or

$$\Phi(A_k) \leqslant C_i \leqslant D_j \leqslant \Phi(B).$$

Since $(w(A_k), w(B)) < (w(A), w(B))$, it thereby follows that

$$\overline{A} \leqslant \overline{A_k} \leqslant \overline{B}.$$

Now we can turn to the proof of Theorem 2. Let $A, B \in \mathfrak{G}$. If $\overline{A} = \overline{B}$, then by ($c$), we have

$$\overline{\Phi(A)} \leqslant \overline{\Phi(B)} \leqslant \overline{\Phi(A)},$$

that is, $\overline{\Phi(A)} = \overline{\Phi(B)}$ .If $\overline{A} \leqslant \overline{B}$, then, by the same criterion, $\overline{\Phi(A)} \leqslant \overline{\Phi(B)}$. Hence Φ induces an isotone mapping $\overline{\Phi}$ of the lattice G onto a sublattice U of F, generated by the elements $u_1, \ldots, u_m$. If $\overline{\Phi}(A) \leqslant \overline{\Phi}(B)$, then, by (*f*), $\overline{A} \leqslant \overline{B}$. In particular, if $\overline{\Phi}(A) \leqslant \overline{\Phi}(B) \leqslant \overline{\Phi}(A)$, then $\overline{A} = \overline{B}$. Consequently, $\overline{\Phi}$ is an isomorphism, and the theorem is proved.

THEOREM 3. (Whitman). *FL* (3) *contains FL* (*n*) *for all n.*

Proof. Let $\{x_1, x_2, x_3\}$ be the generators of the free lattice $FL(3)$. Put

$$u_1 \equiv x_1(x_2 + x_3) + x_2(x_1 + x_3),$$

$$u_2 \equiv x_1(x_2 + x_3) + x_3(x_1 + x_2),$$

$$u_3 \equiv (x_1 + x_2x_3)(x_2 + x_1x_3),$$

and

$$u_4 \equiv (x_1 + x_2x_3)(x_3 + x_1x_2).$$

Consider the lattices S and T of Figure 6. Construct the homomorphisms φ_i $(i = 1, 2, 3, 4)$ of $FL(3)$ into S and T, satisfying the following conditions: (*a*) $\varphi_1(x_1) = y_1$, $\varphi_1(x_2) = y_2$, $\varphi_1(x_3) = y_3$; (*b*) $\varphi_2(x_1) = y_1$, $\varphi_2(x_2) = y_3$, $\varphi_2(x_3) = y_2$; (*c*) $\varphi_3(x_1) = z_1$, $\varphi_3(x_2) = z_3$, $\varphi_3(x_3) = z_2$; (*d*) $\varphi_4(x_1) = z_1$, $\varphi_4(x_2) = z_2$, $\varphi_4(x_3) = z_3$. Then

$$\varphi_1(u_1) = y_1d + y_2b = 0 + a = a,$$

$$\varphi_1(u_2) = y_1d + y_31 = 0 + y_3 = y_3,$$

$$\varphi_1(u_3) = (y_1 + 0)(y_2 + 0) = 0,$$

$$\varphi_1(u_4) = (y_1 + 0)(y_3 + 0) = 0.$$

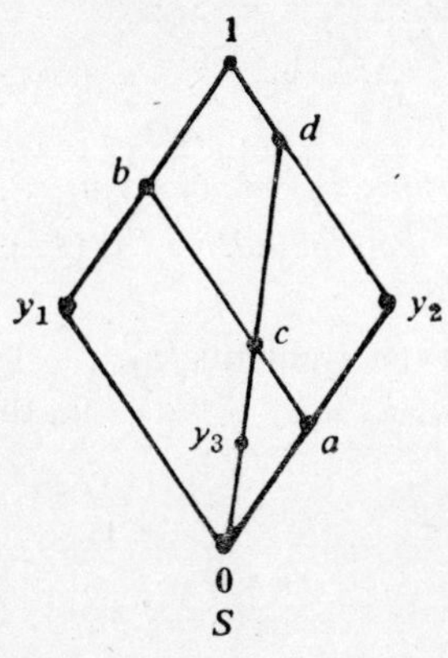

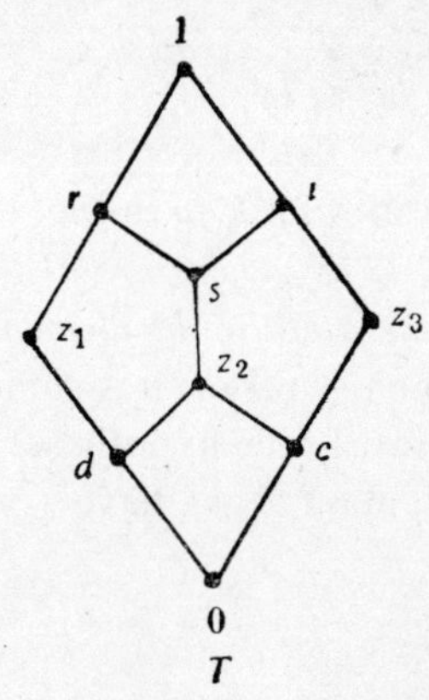

Figure 6.

If $u_1 \leqslant u_2 + u_3 + u_4$, then

$$a = \varphi_1(u_1) \leqslant \varphi_1(u_2) + \varphi_1(u_3) + \varphi_1(u_4) \leqslant y_3 + 0 + 0 = y_3,$$

which is false. Similarly, by means of the homomorphisms φ_2, φ_3 and φ_4, we reach a contradiction from the assumptions that

$$u_2 \leqslant u_1 + u_3 + u_4, \quad u_3 \leqslant u_1 + u_2 + u_4 \quad \text{and} \quad u_4 \leqslant u_1 + u_2 + u_3.$$

By duality we see that the inequalities $u_1 \geqslant u_2u_3u_4$, $u_2 \geqslant u_1u_3u_4$, $u_3 \geqslant u_1u_2u_4$ and $u_4 \geqslant u_1u_2u_3$, are impossible. Thus, the elements u_1, u_2, u_3, u_4 satisfy the hypotheses of Theorem 2 and, consequently, generate the lattice FL (4). The assertion that follows is immediate from this conclusion.

LEMMA 1. *FL (3) contains FL (4).*

LEMMA 2. *If $\{y_1, \ldots, y_m\}$ are the generators of the free lattice $F = FL(m)$ and a is an element of the sublattice G of F, generated by the elements $\{y_1, \ldots, y_{m-1}\}$, then $a \leqslant y_1 + \cdots + y_{m-1}$.*

Indeed, the inequality in hand holds, if the weight of the element a is zero. Otherwise, we have $a = b + c$ or $a = bc$, where b and c have a least weight. By induction, we obtain

$$a = b + c \leqslant (y_1 + \cdots + y_{m-1}) + (y_1 + \cdots + y_{m-1}) = y_1 + \cdots + y_{m-1},$$

or

$$a = bc \leqslant (y_1 + \cdots + y_{m-1})(y_1 + \cdots + y_{m-1}) = y_1 + \cdots + y_{m-1}.$$

LEMMA 3. *If in the notations of Lemma 2, we have the elements $u_1, \ldots, u_s \in G$ as generators of the free lattice $FL(s)$, then the lattice $FL(s+1)$ is generated by the elements $\{u_1, \ldots, u_s, y_m\}$.*

For the proof, it is sufficient to establish that the elements $\{u_1, \ldots, u_s, y_m\}$ satisfy the hypotheses of Theorem 2. If $y_m \leqslant u_1 + \cdots + u_s$, then, by Lemma 2, we have

$$y_m \leqslant u_1 + \cdots + u_s \leqslant y_1 + \cdots + y_{m-1},$$

contradicting Theorem 2. But, then, if

$$u_i \leqslant u_1 + \cdots + u_{i-1} + u_{i+1} + \cdots + u_s + y_m,$$

we consider the homomorphism ψ of the lattice F into itself, such that

$\psi(y_j) = y_j$ $(j = 1, \ldots, m - 1)$ and $\psi(y_m) = u_1 \ldots u_s$. We get

$$u_i = \psi(u_i) \leqslant u_1 + \cdots + u_{i-1} + u_{i+1} + \cdots + u_s + u_1 \ldots u_s$$

$$\leqslant u_1 + \cdots + u_{i-1} + u_{i+1} \cdots + u_m,$$

which again contradicts Theorem 2. This shows that property (5.5) of Theorem 2 is valid and dually so is property (5.6).

LEMMA 4. *If $m \geqslant 3$, then $FL(m)$ contains $FL(m + 1)$.*

Indeed, for $m = 3$ this has been proved in Lemma 1. For $m > 3$, we employ the notations of Lemma 2 and the induction principle to select in the lattice G, the elements $u_1, \ldots, u_m$ generating $FL(m)$. In view of Lemma 3, the lattice $FL(m + 1)$ is generated by the elements $u_1, \ldots, u_m, y_m$.

We return to the proof of the theorem and note that for $n = 4$, everything is proved in Lemma 1. However, if $FL(3)$ contains $FL(m)$, then it contains $FL(m + 1)$ in view of Lemma 4.

As a trivial corollary to Theorem 3, we have the next theorem.

THEOREM 4. *If $n \geqslant 3$, then the lattice $FL(n)$ is infinite.*

Thus, $FL(1)$ and $FL(2)$ are the only finite free lattices. It is found that among the free lattices there are no other complete lattices.

THEOREM 5. *Let F be the free extension of a poset P, containing the elements u_1, u_2, u_3, such that for any permutation (i, j, k) of the elements $\{1, 2, 3\}$ the following hold:*

(i) *If $x \equiv u_i$, $y \equiv u_j$ and $z \equiv u_k$, then $x \nleqslant_F y + z$ and $x \ngeqslant_F yz$;*

(ii) *If $x \equiv u_i$, $y \equiv u_j$, $z \equiv u_k$, $a \in P$ and if there exists a value*

$$v(x + y(z + x(y + za))),$$

then

$$a \neq x + y(z + x(y + za));$$

(iii) *if $x \equiv u_i$, $y \equiv u_j$, $z \equiv u_k$, $a \in P$ and if there exists a value*

$$v(x(y + z(x + y(z + a)))),$$

then

$$a \neq x(y + z(x + y(z + a))).$$

Then F is not a complete lattice. Moreover, the properties (ii) *and* (iii) *remain valid for any* $a \in F$.

Proof. By Theorem 1, the lattice F is a quotient set of the set of words, constructed at the beginning of the present chapter. The elements of the poset P may be identified with the corresponding words of zero weight. As before, we shall use the symbols $=$, $\equiv$, the value of the word $v(A)$ and its weight $w(A)$. The word A is said to be *reduced* if for its $\cup$- and $\cap$-canonical expansions

$$A \equiv A_1' + \cdots + A_m',$$

and

$$A \equiv A_1'' \cdots A_n'',$$

for every i and A_0, where A_0 is either the void symbol or $w(A_0) < w(A_i)$,

$$A \neq A_1' + \cdots + A_{i-1}' + A_0 + A_{i+1}' + \cdots + A_m',$$

and

$$A \neq A_1'' \ldots A_{i-1}'' A_0 A_{i+1}'' \ldots A_n''$$

hold. The word is said to be *completely reduced*, if the word itself and all its subwords are reduced. A completely reduced word A is called *minimal* if $A = B$ implies $w(A) \leqslant w(B)$. A word is called *completely minimal*, if the word itself and all its subwords are minimal.

LEMMA 1. *For every word B, there is a completely minimal word A, such that* $A = B$.

In fact, a word of weight zero is, evidently, completely minimal. Let

B be an arbitrary word of nonzero weight. Among the words equal to B in the lattice F, we mark the word C of a least weight. If $w(C) = 0$, then we can put $A = C$. Otherwise, we have, say $C \equiv C' C''$, where $w(C')$, $w(C'') < w(C)$. By the induction hypothesis, we can find completely minimal words A' and A'' such that $A' = C'$ and $A'' = C''$. It is easy to see that the word $A \equiv A'A''$ satisfies the hypotheses of the lemma.

LEMMA 2. *If A is a minimal word, if $w(A) \neq 0$ and $A = B$, then the value $v(B)$ does not exist.*

Indeed, if $v(B)$ exists, then

$$A \leqslant B \underset{1}{\leqslant} v(B) \underset{1}{\leqslant} B \leqslant A.$$

Consequently, $A = v(B)$. But $w(v(B)) = 0 < w(A)$, which contradicts the minimality of A.

LEMMA 3. *If A is a minimal word, $w(A) \neq 0$ and $A = a + B$, where $a \in P$ and $a \nleqslant B$, then $A \equiv A_1 + A_2$, where $A_1 = a$ and $A_2 = B$ or $A_1 = B$ and $A_2 = a$.*

For proof, we write the $\cup$-canonical representation

$$A \equiv A_1 + \cdots + A_m.$$

Lemma 2 allows us to apply Proposition 10 to the inequality

$$a \leqslant A_1 + \cdots + A_m,$$

which yields $a \leqslant A_1$ for suitably chosen indices. By the same criterion, from Proposition 10 or 12, applied to the inequality

$$A_1 \leqslant a + B,$$

it follows that either $A_1 \leqslant a$, or $A_1 \leqslant B$, or $E \leqslant a + B$, for some word E, such that $E \geqslant A_1$ and $w(E) < w(A_1)$. However, the second supposition leads to the following inequality in contradiction to the hypothesis

$$a \leqslant A_1 \leqslant B.$$

From the third supposition, it follows that

$$A \leqslant E + A_2 + \cdots + A_m \leqslant a + B + A_2 + \cdots + A_m \leqslant A,$$

contradicting the reducibility of A. Thus $a \leqslant A_1 \leqslant a$, that is, $A_1 = a$. In view of the minimality of A, it thereby follows that $A \not\equiv A_1$. Hence, applying Proposition 10 or 12 to the inequality

$$A_2 \leqslant a + B,$$

we have either $A_2 \leqslant B$, or $A_2 \leqslant a$, or $E \leqslant a + B$, where $E \geqslant A_2$ and $w(E) < w(A_2)$. Since $A_1 = a$, the second supposition leads to the equality

$$A = A_1 + A_3 + \cdots + A_m, \tag{5.7}$$

contradicting the reducibility of A. This is also contradicted by the relation

$$A \leqslant A_1 + E + A_3 + \cdots + A_m \leqslant A_1 + a + B + A_3 + \cdots + A_m \leqslant A,$$

which follows from the third supposition. Thus, $A_2 \leqslant B$. Finally, applying Proposition 10 or 12 to the inequality

$$B \leqslant A_1 + \cdots + A_m,$$

for some number i, we get $B \leqslant A_i$. If $i \neq 2$, then $A_2 \leqslant B \leqslant A_i$, whence

$$A = A_1 + A_3 + \cdots + A_m,$$

violating the reducibility of A. Consequently, $A_2 \leqslant B \leqslant A_2$, that is, $A_2 = B$. However, then

$$A = A_1 + A_2,$$

and in view of the reducibility of A, we have $A \equiv A_1 + A_2$, where $A_1 = a$, $A_2 = B$.

Since, during the proof the indices have been interchanged, it is possible that $A_1 = B$ and $A_2 = a$.

The next lemma is proved by duality.

LEMMA 4. *If A is a minimal word, $w(A) \neq 0$ and $A = aB$, where $a \in P$ and $a \not\geqslant B$, then $A \equiv A_1A_2$, where $A_1 = a$ and $A_2 = B$ or $A_1 = B$ and $A_2 = a$.*

LEMMA 5. *If $a \in P$, then*

$$a \neq x + y(z + x(y + za)),$$

and

$$a \neq x(y + z(x + y(z + a))).$$

For proof, we agree that

$$a = x + y(z + x(y + za)),$$

and put

$$U \equiv z + x(y + za).$$

If there exists a value $v(yU)$, then by Proposition 13, we have

$$a = \sup_P \{x, v(yU)\}.$$

Since $\mathfrak{A}$ contains all summable pairs of elements in P, it thereby implies the existence of the value

$$v(x + yU),$$

in contradiction to condition (ii). However, if the value $v(yU)$ does not exist, then, applying Proposition 11 to the inequality $yU \leqslant a$ we obtain

$$y \leqslant a = x + yU \leqslant x + U \leqslant x + z,$$

or

$$z \leqslant U \leqslant a \leqslant x + y.$$

However, each of these inequalities contradicts condition (i).

The proof of the second inequality set forth in Lemma 5, follows by duality.

LEMMA 6. *If $A \in F$, then*

$$A \neq x + y(z + x(y + zA)),$$

and

$$A \neq x(y + z(x + y(z + A))).$$

For the proof, assume that

$$A = x + y(z + x(y + zA)).$$

In view of Lemma 5, $w(A) \neq 0$. Lemma 1 allows us to treat A as a completely minimal word. Put

$$C \equiv z + x(y + zA),$$

and

$$D \equiv y + zA.$$

Consider the equality

$$A = x + yC.$$

If $x \leqslant yC$, then $x \leqslant yC \leqslant y + z$, violating condition (i). Hence, applying Lemma 3, we get

$$A \equiv A' + A_1,$$

or

$$A \equiv A_1 + A',$$

where $A' = x$ and $A_1 = yC$. If $w(A_1) = 0$, that is, $A_1 \in P$, then the equality

$$A_1 = y(z + x(y + zA)) = y(z + x(y + z(x + A_1)))$$

contradicts Lemma 5. However, if $w(A_1) \neq 0$, then we note that the inequality $y \geqslant C$ leads to the relation

$$x + y \geqslant y \geqslant C \geqslant z,$$

contradicting condition (i), which allows us to apply Lemma 4 to the equality $A_1 = yC$. Consequently, $A_1 \equiv A''A_2$, or $A_1 \equiv A_2A''$, where $A'' = y$ and $A_2 = C = z + xD$. As above, the assumption $w(A_2) = 0$ leads to the equality

$$A_2 = z + x(y + zA) = z + x(y + z(x + yC))$$
$$= z + x(y + z(x + yA_2)),$$

contradicting Lemma 5. However, if $w(A_2) \neq 0$, then we note that the inequality $z \leqslant xD$ leads to the condition (i) being contradicted by the relation

$$z \leqslant xD \leqslant x + y,$$

by applying Lemma 3 to the equality

$$A_2 = z + xD.$$

Therefore,

$$A_2 \equiv A''' + A_3,$$

or

$$A_2 \equiv A_3 + A''',$$

where $A''' = z$ and $A_3 = xD$. If $w(A_3) = 0$, then

$$A_3 = x(y + zA) = x(y + z(x + yC)) = x(y + z(x + y(z + x D)))$$
$$= x(y + z(x + y(z + A_3))),$$

contradicting Lemma 5. If $x \geqslant D$, then

$$x + z \geqslant D \geqslant y,$$

which contradicts condition (i). Consequently, if we apply Lemma 4 to the equality $A_3 = xD$, we obtain either $A_3 \equiv A^{IV}A_4$, or $A_3 \equiv A_4A^{IV}$, where $A^{IV} = x$ and $A_4 = D = y + zA$. If $w(A_4) = 0$, then the equality

$$A_4 = y + zA = y + z(x + y(z + xA_4)),$$

violating Lemma 5. If $y \leqslant zA$, then

$$y \leqslant zA \leqslant x + z,$$

contradicting condition (i). Consequently, we can apply Lemma 3 to the equality $A_4 = y + zA$, yielding either $A_4 \equiv A^V + A_5$, or $A_4 \equiv A_5 + A^V$, where $A^V = y$ and $A_5 = zA$. If $w(A_5) = 0$, then, we have

$$A_5 = zA = z(x + y(z + x(y + A_5))),$$

violating Lemma 5. If $z \geqslant A$, then

$$y + z \geqslant A \geqslant x,$$

which violates condition (i). As above, this allows us to apply Lemma 4 to the equality $A_5 = zA$. Consequently, $A_5 \equiv A^{VI}A_6$, or $A_5 \equiv A_6A^{VI}$, where $A_6 = A$. However,

$$w(A_6) < w(A_5) < w(A_4) < w(A_3) < w(A_2) < w(A_1) < w(A),$$

contradicting the minimality of A.

The second inequality is proved on similar lines.

For the proof of theorem, note that by Lemma 6, an isotone map φ of the lattice F into itself, defined by the equality

$$\varphi(A) = x + y(z + x(y + zA)),$$

has no fixed point. Hence, by Theorem 3.2, the lattice F can not be complete. The validity of properties (ii) and (iii) for the elements of F is established in Lemma 6.

From Theorems 2 and 5 the next conclusion is immediate.

THEOREM 6. *If $n \geqslant 3$, then the lattice $FL(n)$ is not complete.*

Naturally, we now desire to define in the class of all complete lattices the analogue of free lattices. For a trivial poset P, the corresponding definition must be as follows: a complete lattice F is called the *free complete lattice* with a system of free generators P, if

(1) $P \subseteq F$;

(2) F is completely generated by P;

(3) for every mapping φ of P into an arbitrary complete lattice L, there is a complete homomorphism ψ of F into L, such that $\psi(x) = \varphi(x)$ for all $x \in P$.

Note that the existence of the free complete lattice does not follow from the theory of universal algebra, since in a complete lattice we have to consider infinitary operations.

The next theorem facilitates the solution of the existence problem for free latti ces.

THEOREM 7. *Let L be a lattice and U a sublattice of L, such that* $\sup_L U$ *does not exist. Then there exists a lattice F, containing L as a sublattice together with $u =$* $\sup_F U$, *which is completely generated by the set L.*

For the proof, adjoin to L a new element u and on the obtained set P define the relation:

$$a \trianglelefteq b \underset{\text{def}}{=} \boxed{\begin{array}{l} \text{(i) } a, b \in L \text{ and } a \leqslant_L b; \\ \qquad\qquad \text{or} \\ \text{(ii) } a = u, b \in L \text{ and } x \leqslant_L b \\ \qquad \text{for all } x \in U; \\ \qquad\qquad \text{or} \\ \text{(iii) } a \in L, b = u \text{ and } a \leqslant_L x_0 \\ \qquad \text{for some } x_0 \in U; \\ \qquad\qquad \text{or} \\ \text{(iv) } a = b = u. \end{array}}$$

The reflexivity of $\trianglelefteq$ is obvious. If $a \trianglelefteq b$ and $b \trianglelefteq a$, then the applicability of rules is possible according to the following combination scheme: (*a*) (i) and (i); (*b*) (ii) and (iii); (*c*) (iii) and (ii); (*d*) (iv) and (iv). In cases (*a*) and (*d*), it is clear that $a = b$. In case (*b*) we have $a = u$ and $b \in L$ and also $b \leqslant x_0$ for some $x_0 \in U$ and $x \leqslant b$ for all $x \in U$. Since $\sup U$ does not exist, we have $x_0 < x_1$ for some $x_1 \in U$; therefore, $x_1 \leqslant \leqslant b \leqslant x_0 < x_1$, which is impossible. In case (*c*) we must have $a \in L$ and $b = u$ and also $x \leqslant a$ for all $x \in U$ and $a \leqslant x_0$ for some $x_0 \in U$. As above, we arrive at a contradiction, getting $x_1 \leqslant a \leqslant x_0 < x_1$. Thus $\trianglelefteq$ is an antisymmetric relation. If $a \trianglelefteq b$ and $b \trianglelefteq c$, then the applicability of rules according to the following combinations is possible: (*a*) (i) and (i); (*b*) (i) and (iii); (*c*) (iii) and (ii); (*d*) (iii) and (iv); (*e*) (ii) and (i); (*f*) (ii) and (iii); (*g*) (iv) and (ii); (*h*) (iv) and (iv). In cases (*a*), (*d*), (*f*), (*g*) and (*h*) the relation $a \trianglelefteq c$ evidently holds. In case (*b*) we have $c = u$ and $a \leqslant b \leqslant x_0$ for some $x_0 \in U$, whence, $a \trianglelefteq c$ by (iii). In case (*e*) we have $a = u$ and $x \leqslant b \leqslant c$ for all $x \in U$, whence $a \trianglelefteq c$ by (ii). In case (*c*), $b = u$, $a \leqslant x_0$ for some $x_0 \in U$ and $x \leqslant c$ for all $x \in U$. In particular, $a \leqslant x_0 \leqslant c$ and, consequently, $a \trianglelefteq c$ by (i) Thus, P is found to be a poset. Let F be its free extension. From properties (i)–(iv) of the $(\mathfrak{A}, \mathfrak{B})$-free extension, it follows that F contains L as a sublattice and is generated by P. The proof will be complete, if it is established that $u = \sup_F U$. Clearly, $u \geqslant_F x$ for all $x \in U$. If $a \in F$ and $a \geqslant_F x$ for all $x \in U$, then in case the weight of a is zero, by rule (ii) we have $a \trianglelefteq u$ and, consequently, $a \geqslant_F u$. Otherwise, we have a reduced $\cap$-canonical [$\cup$-canonical] representation

$$a \equiv b_1 b_2 \ldots b_s \qquad [a \equiv b_1 + \cdots + b_s],$$

where all b_i $(i = 1, \ldots, s)$ have a smaller weight. In the first case, for every $x \in U$, $x \leqslant_F a \leqslant_F b_i$ $(i = 1, \ldots, s)$ holds. Taking note of the induction hypothesis, we get $u \leqslant_F b_i$ and, consequently, $u \leqslant_F a$. For the second case, assume that for every b, there is an element $x_i \in U$ such that $x_i \nleqslant_F b_i$. However, $x = x_1 + \cdots + x_s \in U$. Consequently, $x \leqslant_F a$ and by Proposition 10, $x \leqslant_F b_j$ for some j. Hence

$$x_j \leqslant_F x \leqslant_F b_j,$$

contradicting the choice of the element x_j. Hence, for some i, the inequality $x \leqslant_F b_i$ holds for all $x \in U$. By induction, we get

$$u \leqslant_F b_i \leqslant_F a.$$

This shows that $u = \sup_F U$.

THEOREM 8. *For any set* $\mathfrak{M}$ *there exists a complete lattice L, completely generated by three elements, such that the power of L is not less than the power of* $\mathfrak{M}$.

Proof. By using Theorem 2.3, we can make $\mathfrak{M}$ a well ordered set. Let $F_0 = FL\,(3)$ and let F_0 be free lattice with generators u_1, u_2, u_3. Assume that for each $\beta < \alpha$ $(\alpha, \beta \in \mathfrak{M})$, a lattice F_β has been constructed, satisfying the following:

(*a*) F_β is completely generated by the elements u_1, u_2, u_3:

(*b*) if $\beta < \gamma < \alpha$, then $F_\beta \underset{\neq}{\subset} F_\gamma$;

(*c*) F_β has properties (i)–(iii) of Theorem 5;

(*d*) F_β is not a complete lattice.

By Theorems 2, 5 and 6, F_0 has properties (*a*), (*c*) and (*d*). However, property (*b*) has no meaning for it. If α is a limit ordinal, then we put $F_\alpha = \bigcup_{\beta<\alpha} F_\beta$. By property (ii) of Theorem 5, the mapping φ, defined by

$$\varphi\,(a) = u_1 + u_2\,(u_3 + u_1\,(u_2 + u_3 a)),$$

is an isotone map of the lattice F_α into itself without a fixed point. Hence, from Theorem 3.2, it follows that F_α has property (*d*). The validity of properties (*a*)–(*c*) is obvious. If $\alpha - 1$ exists, then by Theorem 4.12, either $\inf_{F_{\alpha-1}} F_{\alpha-1}$ does not exist, or $F_{\alpha-1}$ contains a sublattice U, such that $\sup_{F_{\alpha-1}} U$ does not exist. Applying to $F_{\alpha-1}$ Theorem 7 or its dual, we get the lattice F_α. This, evidently, has properties (*a*) and (*b*). From Theorem 5, the validity of properties (*c*) and (*d*) is immediate. Put $F = \bigcup_{\alpha \in \mathfrak{M}} F_\alpha$. By property (*b*) the power of F is not less than that of $\mathfrak{M}$. In view of (*a*) F is completely generated by three elements. By Theorem 3.7, both these properties are evidently preserved for the completion by cuts $\bar{F}$ of the lattice F, and the proof is complete.

THEOREM 9. (Crawley–Dean). *There exists no free complete lattice with three generators.*

Proof. Let F be a free complete lattice with three generators. From Theorem 8 and the well known results of set theory (see, for example, P. S. Alexanderov, *Introduction to General Theory of Sets and Functions*, Gostekhizdat, 1948, p. 42, Theorem 15) follows the existence of a complete lattice L, completely generated by three elements and having a power greater than the power of the lattice F. However, then, F cannot be mapped onto L, contradicting property (iii) of the definition of free complete lattices.

Exercises

1. Free and completely free extensions of a poset P coincide, iff P is trivial.
2. Any lattice is a homomorphic image of a free lattice.
3. $FL(n)$ is a latttice with zero and unit element.
4. A free extension of an infinite trivial poset has neither zero nor unit element.
5. The $(\mathfrak{A}, \mathfrak{B})$-free extension of a poset with zero (unit) has zero (unit).
6. $FL(3)$ contains $FL(\aleph_0)$.
7. A free extension F of a poset P is defined uniquely to within isomorphism over P, that is, for any other such modular extension F', there is an isomorphism F onto F', leaving all elements in P fixed.

6

MODULAR LATTICES

A lattice L is called a *Dedekind* (or *modular*) lattice if for any $a, b, c \in L$, where $a \leqslant c$, the modular law

$$(a + b)\, c = a + bc$$

holds. Note that the identity dual to the modular law has the form $ab + c = a\,(b + c)$, where $a \geqslant c$, that is, it again forms a modular law. Thus, the modular law is self-dual. Consequently, a poset dual to a modular lattice, is again a modular lattice. Hence, the duality principle applies to modular lattices.

An important example of a modular lattice is the lattice of all subspaces of a linear space. Indeed, if A, B, C are the subspaces and $A \subseteq C$, then evidently $(A + B) \cap C \supseteq A + B \cap C$. Conversely, if the vector c lies in $(A + B) \cap C$, then, $c = a + b$, where $a \in A, b \in B$. Therefore $b = c - a \in B \cap C$, that is, $C \in A + B \cap C$ and, consequently, $(A + B) \cap C \subseteq A + B \cap C$. Similarly, it is verified that the lattices of normal subgroups of an arbitrary group, the ideals of a ring, the submodules of a module are modular. By contrast, the lattice of all subgroups and the lattice of all equivalences are not necessarily modular. The lattice of linear subvarieties of an affine space is also not modular.

THEOREM 1. *The following properties of a lattice L are equivalent:*

(i) *L is modular;*

(ii) $a(ab + c) = ab + ac$ *for any* $a, b, c \in L$;

(iii) *if* $a \geqslant b$ *and if for some* $c \in L$, $a + c = b + c$ *and* $ac = bc$ *hold, then* $a = b$.

Proof. Assume that property (i) holds. Then for any $a, b, c \in L$ we have $ab \leqslant a$ and, consequently

$$a(ab + c) = (ab + c)a = ab + ac.$$

If in addition, the conditions of property (iii) are satisfied, then

$$a = a(a + c) = a(b + c) = b + ac = b + bc = b.$$

Thus, the implications (i) $\Rightarrow$ (ii) and (i) $\Rightarrow$ (iii) hold. If (ii) is satisfied and $a \leqslant c$, then we have

$$(a + b)c = (ac + b)c = ac + bc = a + bc,$$

that is, (ii) $\Rightarrow$ (i). Assume, finally, that property (iii) holds. If $a, b, c \in L$ and $a \leqslant c$, then

$$a + b \leqslant (a + b)c + b \leqslant a + b,$$

and

$$bc \leqslant (a + bc)b \leqslant (c + bc)b = bc.$$

Therefore

$$(a + b)c + b = a + b,$$

$$(a + bc) + b = a + b,$$

$$(a + b)cb = bc,$$

and

$$(a + bc)b = bc.$$

Since

$$(a + b)\, c \geqslant a + bc,$$

then, by applying (iii), we obtain

$$(a + b)\, c = a + bc.$$

Thus, the implication (iii) ⇒ (i) also holds.

From the theorem proved, it is immediate that the modular lattices form a variety of universal algebra.

THEOREM 2. (Cancellation Rule). *If a, b, c are elements of a modular lattice and $(a + b)\, c = 0$, then $a\,(b + c) = ab$.*

Proof. Taking note of the modular law we obtain

$$a\,(b + c) = a\,(a + b)\,(b + c) = a\,(b + (a + b)\, c) = ab.$$

THEOREM 3. *A complemented modular lattice is relatively complemented.*

Proof. Let $a \leqslant x \leqslant b$, $x + y = 1$ and let $xy = 0$. Then, by the modular law, we get

$$x + (a + yb) = x + yb = (x + y)\, b = b,$$

and

$$x\,(a + yb) = a + xyb = a,$$

giving the required proof.

In a modular lattice with zero, we can define a natural extension of the well known notion of independence from linear algebra. Thus, the elements $a_1, \ldots, a_n$ of a modular lattice L with zero are called *independent*, if

$$(a_1 + \cdots + a_{i-1} + a_{i+1} + \cdots + a_n)\, a_i = 0$$

for all $i = 1, \ldots, n$. From the definition the next theorem is immediate.

THEOREM 4. *If $a_i \leqslant b_i$ $(i = 1, \ldots, n)$ and $b_1, \ldots, b_n$ are independent, then $a_1, \ldots, a_n$ are also independent.*

If $a_1, \ldots, a_n$ are independent elements, then their sum is called *direct* and written as

$$a_1 \dotplus \cdots \dotplus a_n.$$

THEOREM 5. *If $a_1, \ldots, a_n$ are such that*

$$(a_1 + \cdots + a_{i-1})\, a_i = 0$$

for all $i = 1, \ldots, n$, then they are independent.

Proof. For every fixed $j = 1, \ldots, n$, we shall prove the inequalities

$$(a_1 + \cdots + a_{j-1} + a_{j+1} + \cdots + a_l)\, a_j = 0$$

where $l = j + 1, \ldots, n$. Assume that this is true for $l - 1$. Then, we note that

$$[(a_1 + \cdots + a_{j-1} + a_{j+1} + \cdots + a_{l-1} + a_j]\, a_l = 0,$$

and applying the cancellation law and induction hypothesis, we obtain

$$\begin{aligned}(a_1 + \cdots + a_{j-1} + a_{j+1} + \cdots + a_l)\, a_j \\ = [(a_1 + \cdots + a_{j-1} + a_{j+1} + \cdots + a_{l-1}) + a_l]\, a_j \\ = (a_1 + \cdots + a_{j-1} + a_{j+1} + \cdots + a_{l-1})\, a_j = 0.\end{aligned}$$

We complete the proof by noting that the hypothesis of the theorem supplies the basis for the induction.

THEOREM 6. *If $a_1, \ldots, a_n$ are independent and I, J are some subsets of $\{1, \ldots, n\}$, then*

$$\Big(\sum_{i \in I} a_i\Big)\Big(\sum_{j \in J} a_j\Big) = \begin{cases} 0, & \text{if } I \cap J \text{ is void,} \\ \displaystyle\sum_{k \in I \cap J} a_k, & \text{otherwise.} \end{cases}$$

Proof. If $J \subseteq I$ or J consists of one element, then assertion is trivial. Further, if $I \cap J$ is void and $j_0 \in J$ then, by the induction hypothesis, we obtain

$$\Big(\sum_{i \in I} a_i + a_{j_0}\Big)\Big(\sum_{j \in J \setminus j_0} a_j\Big) = 0,$$

whence, by the cancellation law, it follows that

$$\sum_{i \in I} a_i \cdot \sum_{j \in J} a_j = \Big(\sum_{i \in I} a_i\Big)\Big(a_{j_0} + \sum_{j \in J \setminus j_0} a_j\Big) = \sum_{i \in I} a_i a_{j_0} = 0.$$

However, if there exists $j_0 \in I \cap J$, then, taking note of the modular law and the induction hypothesis, we obtain

$$\sum_{i \in I} a_i \cdot \sum_{j \in J} a_j = \Big(\sum_{i \in I} a_i\Big)\Big(a_{j_0} + \sum_{j \in J \setminus j_0} a_j\Big)$$

$$= a_{j_0} + \sum_{k \in I \cap J \setminus j_0} a_k = \sum_{k \in I \cap J} a_k.$$

THEOREM 7. *If $a_1, \ldots, a_n$ are independent and*

$$a_i = a_{i1} \dotplus \cdots \dotplus a_{ik_i} \qquad (i = 1, \ldots, n),$$

then the elements

$$a_{11}, \ldots, a_{1k_1}, a_{21}, \ldots, a_{2k_2}, \ldots, a_{n1}, \ldots, a_{nk_n}$$

are independent.

Proof. Since

$$(a_1 + \cdots + a_{i-1})(a_{i1} + \cdots + a_{ij}) \leqslant (a_1 + \cdots + a_{i-1})\, a_i = 0,$$

taking note of the cancellation law, we get

$$(a_1 + \cdots + a_{i-1} + a_{i1} + \cdots + a_{ij-1})\, a_{ij}$$

$$= (a_{i1} + \ldots + a_{ij-1})\, a_{ij} = 0.$$

Now it remains only to apply Theorem 5.

THEOREM 8. *If* $a_1, \ldots, a_n$ *are independent elements of a modular lattice L with zero* 0, *then the sublattice H of L, generated by the intervals* $[0, a_i]$ $(i = 1, \ldots, n)$ *is isomorphic with the direct product*

$$[0, a_1] \times [0, a_2] \times \ldots \times [0, a_n].$$

Proof. Let us first assume that $n = 2$, and consider the set M of all elements of the form $x_1 + x_2$, where $x_1 \leqslant a_1$ and $x_2 \leqslant a_2$. Clearly, this set is closed under addition. If $x_1 + x_2$ and $y_1 + y_2 \in M$, then

$$(x_1 + y_1)(x_2 + y_2) \leqslant a_1 a_2 = 0.$$

Applying the cancellation law, we get

$$x_1 (y_1 + x_2 + y_2) = x_1 y_1.$$

Similarly, it is verified that

$$x_2 (y_1 + x_1 + y_2) = x_2 y_2.$$

From these equalities, with the aid of the modular law, it is deduced that

$$\begin{aligned}(x_1 + x_2)(y_1 + y_2) &= (x_1 + x_2)(y_1 + x_2 + y_2)(y_1 + y_2) \\ &= (x_2 + x_1 (y_1 + x_2 + y_2))(y_1 + y_2) \\ &= (x_2 + x_1 y_1)(y_1 + y_2) \\ &= x_1 y_1 + x_2 (y_1 + y_2)(y_1 + x_1 + y_2) \\ &= x_1 y_1 + x_2 y_2 (y_1 + y_2) = x_1 y_1 + x_2 y_2.\end{aligned}$$

Thus, the set M is closed also under multiplication and, hence, coincides with the sublattice H. Moreover, the last equality together with the obvious equality

$$(x_1 + x_2) + (y_1 + y_2) = (x_1 + y_1) + (x_2 + y_2)$$

shows that

$$\varphi (x, y) = x + y$$

is the lattice homomorphism of the direct product $[0, a_1] \times [0, a_2]$ onto H. If $x_1 + x_2 = y_1 + y_2$, then, since $y_2 a_1 \leqslant a_2 a_1 = 0$ and $x_2 a_1 \leqslant$ $\leqslant a_2 a_1 = 0$, the modular law yields

$$x_1 = x_1 (x_1 + x_2) a_1 = x_1 (y_1 + y_2) a_1 = x_1 (y_1 + y_2 a_1) = x_1 y_1$$
$$= y_1 (x_1 + x_2 a_1) = y_1 (x_1 + x_2) a_1 = y_1 (y_1 + y_2) a_1 = y_1.$$

Similarly, it is established that $x_2 = y_2$. Consequently, φ is an isomorphism. If $n > 2$, then, using the induction hypothesis, we obtain

$$H \cong [0, a_1 + \cdots + a_{n-1}] \times [0, a_n]$$
$$\cong [0, a_1] \times \cdots \times [0, a_{n-1}] \times [0, a_n].$$

THEOREM 9. *If a and b are the elements of a modular lattice L, then the intervals $[ab, a]$ and $[b, a + b]$ are isomorphic. For this isomorphism there exist the mappings*

$$\varphi (x) = x + b \qquad (ab \leqslant x \leqslant a),$$

and

$$\psi (y) = ay \qquad (b \leqslant y \leqslant a + b).$$

Proof. Clearly, φ and ψ are isotone mappings, and

$$\psi\varphi (x) = a (x + b) = x + ab = x,$$

and

$$\varphi\psi (y) = ay + b = (a + b) y = y.$$

If $a \dot{+} c = b \dot{+} c$, then the elements a and b are called *perspective*, and the element c is called an *axis of perspectivity* for a and b. A successive application of the mappings, stated in Theorem 9

$$[0, a] = [ac, a] \to [c, a + c] = [c, b + c] \to [bc, b] = [0, b],$$

enables us to establish an isomorphism between the intervals $[0, a]$ and

$[0, b]$, called a *perspective mapping* with axis c and denoted by $P_{(a \to b\,;\,c)}$. It is easy to see that $P_{(a \to b;\, c)}$ maps the element $x \in [0, a]$ onto the element $y = (x + c)\, b \in [0, b]$.

If L is a lattice, consisting of the void set, points and lines of a projective plane and the plane itself, then the perspective mapping $P_{(a \to b;\, c)}$ establishes a correspondence between the point x on line a to the point y on line b (Figure 7). This also explains the name.

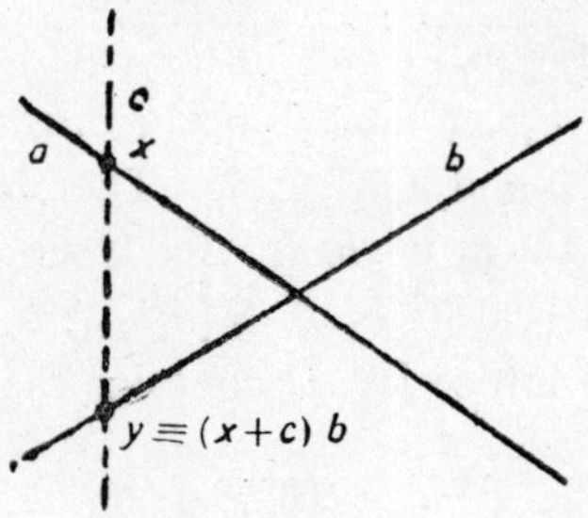

Figure 7.

THEOREM 10. *If a modular lattice L has a composition series of length n, then it does not have a chain containing n + 2 elements.*

Proof. Let $0 = a_0 < a_1 < \ldots < a_n = 1$ be a composition series of the lattice L. Assume that L contains the chain

$$b_0 < b_1 < \ldots < b_{n+1},$$

whose elements are all distinct: Without loss of generality, it can be assumed that $b_0 = 0$. If $n = 1$, then L is a two-element chain and the theorem evidently holds. If $n > 1$, then consider the chain

$$a_1 \leqslant a_1 + b_1 \leqslant \ldots \leqslant a_1 + b_{n+1}.$$

Since this chain belongs to the lattice $[a_1, 1]$ having the composition series

$$a_1 < a_2 < \ldots < a_n = 1,$$

it follows by induction that for some index i, we have

$$a_1 + b_{i-1} = a_1 + b_i.$$

Obviously i can be taken as the least member of such indices. If $i = 1$, that is, $a_1 = a_1 + b_1$, then $0 < b_1 \leqslant a_1$, whence $a_1 = b_1$ in view of $[0, a_1]$ being a simple interval.

Consequently, $[a_1, 1]$ contains the chain $b_1 < \ldots < b_{n+1}$, contradicting the induction hypothesis. Now, let $i > 1$. If

$$a_1 b_{i-1} = a_1 b_i,$$

then property (iii) of Theorem 1 leads to the equality $b_1 = b_{i-1}$ which is false. Otherwise, since $a_1 b_{i-1}$ and $a_1 b_i$ belong to the simple interval $[0, a_1]$, we have $a_1 b_{i-1} = 0$ and $a_1 b_i = a_1$. Therefore, $a_1 \leqslant b_i$ and, consequently, $b_i = a_1 + b_i = a_1 + b_{i-1}$. By Theorem 8, there exists an isomorphism φ of the interval $[0, b_{i-1}] = [a_1 b_{i-1}, b_{i-1}]$ onto $[a_1, a_1 + b_{i-1}] = [a_1, b_i]$. But then the lattice $[a_1, 1]$ contains the chain

$$a_1 = \varphi(b_0) < \varphi(b_1) < \ldots < \varphi(b_{i-1}) = b_i < b_{i+1} < \ldots < b_{n+1},$$

contradicting the induction hypothesis. From Theorem 10, Theorem 11 directly follows.

THEOREM 11. *All composition series of a modular lattice have an identical length.*

Theorem 11 obviously finds applications in the theory of rings, modules and groups.

By a *dimension function* on a lattice L we mean an integer-valued function d, satisfying the following properties:

(1) if $[x, y]$ is a simple interval, then $d(y) = d(x) + 1$;

(2) $d(x + y) + d(xy) = d(x) + d(y)$ for any $x, y \in L$.

If d is a dimension function, then $D \underset{\text{def}}{=} d + m$, m is an integer, and is also a dimension function. Hence, for any given element $a \in L$, a dimension function, if one exists, can always be chosen, such that $d(a) = 0$. A simple induction shows that

$$d(a_1 \dot{+} \cdots \dot{+} a_n) = d(a_1) + \cdots + d(a_n).$$

THEOREM 12. *If an interval* $[a, b]$ *of a modular lattice has a composition* *series, if d is the dimension function and $d(a) = d(b)$ then $a = b$.*

Proof. If $a < a_1 < \ldots < a_{l-1} < b$ is a composition series, then by property (i) we have $d(b) = d(a) + l$, whence $l = 0$ and, consequently, $a = b$.

The chain C in the lattice L is called *bounded*, if its upper and lower cones are not void.

THEOREM 13. *For the lattice L, all of whose bounded chains are finite, the following properties are equivalent:*

(i) *L is modular;*

(ii) *if $a, b \in L$, then the interval $[ab, a]$ is simple iff the interval $[b, a + b]$ is simple;*

(iii) *there exists on L a dimension function d.*

Proof. The implication (i) $\Rightarrow$ (ii) is immediate from Theorem 9. To prove the implication (iii) $\Rightarrow$ (i) we assume that $a, b \in L$, $a \geqslant b$ and for some $c \in L$, we have $a + c = b + c$ and $ac = bc$. Owing to the bounded chains of the lattice L being finite, it follows that the interval $[b, a]$ has composition series. Let l be its length. From the definition of a dimension function it follows that

$$d(b) + l + d(c) = d(a) + d(c) = d(a + c) + d(ac)$$
$$= d(b + c) + d(bc) = d(b) + d(c).$$

Consequently, $l = 0$, that is, $a = b$. It remains to apply Theorem 1. Passing on to the proof of implication (ii) $\Rightarrow$ (iii), we assume that the lattice L has property (ii).

LEMMA 1. *If $a, b \in L$ and $a \leqslant b$, then all composition series of the interval $[a, b]$ have precisely the same length.*

In fact, let l be the least length of the composition series indicated. If $l = 0$ or $l = 1$, then the interval is a one- or two-element chain.

In both cases the validity of the lemma is trivial. Now, let $l > 1$ and let there exist the composition series

$$a = c_0 < c_1 < \ldots < c_l = b,$$

and

$$a = d_0 < d_1 < \ldots < d_m = b.$$

If $c_1 d_1 = c_1$, then $c_1 \leqslant d_1$. We choose the composition series

$$c_1 = f_0 < f_1 < \ldots < f_k = d_1$$

of the interval $[c_1, d_1]$, to obtain the composition series

$$c_1 = f_0 < f_1 < \ldots < f_k = d_1 < \ldots < d_m = b$$

of the interval $[c_1, b]$. Therefore, $l \leqslant m$ and, by the induction hypothesis, $k + m - 1 = l - 1$, whence $k = 0$ and $m = l$. However, if $c_1 d_1 \neq c_1$, then in view of the interval $[a, c_1]$ being simple, we have $c_1 d_1 = a$. Select any composition series

$$c_1 + d_1 = g_0 < g_1 < \ldots < g_n = b$$

of the interval $[c_1 + d_1, b]$. In view of (ii) the chains

$$c_1 < c_1 + d_1 = g_0 < g_1 < \ldots < g_n = b,$$

and

$$d_1 < c_1 + d_1 = g_0 < g_1 < \ldots < g_n = b$$

are found to be composition series of the intervals $[c_1, b]$ and $[d_1, b]$, respectively. Taking note of the induction hypothesis, we get

$$l - 1 = n + 1 = m - 1,$$

and, consequently, $m = l$.

Lemma 1 enables us to denote by the symbol $l(a, b)$ the length of a composition series of the interval $[a, b]$.

LEMMA 2. $l(ab, a) = l(b\ \ a + b)$.

For the proof we put $m = l(ab, a)$. If $m = 0$, then $ab = a$ and hence, $a + b = b$, that is, $l(b, a + b) = 0$. If $m = 1$, then Lemma 2 is immediate from (ii). If $m > 1$, then we choose an element $c \in [ab, a]$ such that $l(ab, c) = 1$ (Figure 8). Then either $bc = c$ or $bc = ab$. However,

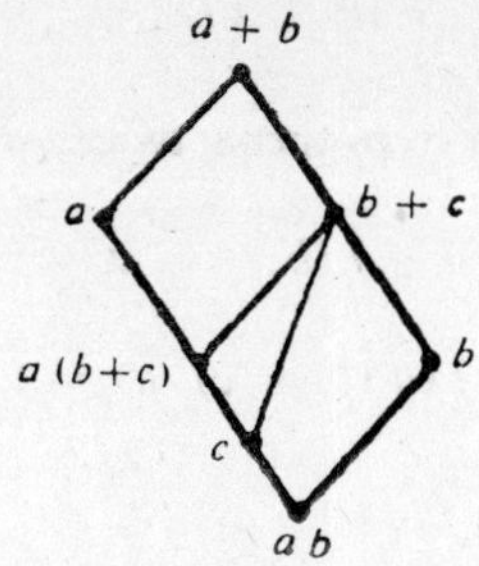

Figure 8.

in the former case we have $b \geqslant c$, and hence, $ab < c \leqslant ab$, which is impossible. Thus $bc = ab$. Furthermore, we have

$$b + c \leqslant b + a(b + c) \leqslant b + c,$$

that is,

$$a(b + c) + b = b + c.$$

It is clear that

$$a(b + c) \cdot b = ab.$$

If $a(b + c) = a$, then

$$b + c = b + c + a = a + b$$

and, since $l(bc, c) = l(ab, c) = 1 < m$,

$$l(b, a + b) = l(b, b + c) = l(ab, c) = 1.$$

From property (ii) it follows that $l(ab, a) = 1$, violating the hypothesis. Hence,

$$l(a(b + c)b, a(b + c)) < m$$

and, by the induction hypothesis, we get

$$l(ab, a(b + c)) = l(a(b + c) \cdot b, a(b + c)) = l(b, a(b + c) + b)$$

$$= l(b, b + c) = l(bc, c) = l(ab, c).$$

Consequently,

$$a(b + c) = c,$$

and applying the induction hypothesis twice, we have

$$\begin{aligned} l(b, a + b) &= l(b, b + c) + l(b + c, a + b) \\ &= l(ab, c) + l(a(b + c), a) = l(ab, c) + l(c, a) \\ &= l(ab, a). \end{aligned}$$

Finally, we choose some fixed point $w \in L$ and put

$$d(x) \underset{\text{def}}{=} l(xw, x) - l(xw, w).$$

If $[x, y]$ is a simple interval, then $xw \leqslant yw \leqslant y$. Hence,

$$\begin{aligned} d(y) &= l(yw, y) - l(yw, w) = l(xw, yw) + l(yw, y) \\ &\quad - [l(xw, yw) + l(yw, w)] = l(xw, y) - l(xw, w) \\ &= l(xw, x) + l(x, y) - l(xw, w) = d(x) + 1. \end{aligned}$$

If $x, y \in L$, then by what has been proved above, it follows that

$$d(x + y) = d(y) + l(y, x + y),$$

and

$$d(xy) = d(x) - l(xy, x).$$

But $l(xy, x) = l(y, x + y)$ by Lemma 2. Consequently

$$d(x + y) + d(xy) = d(x) + d(y).$$

Let L be a modular lattice with zero 0 and unit 1. A non-trivial element $a \in L$ is called *indecomposable*, if it cannot be represented in the form $a = b \dot{+} c$, where $b, c \neq 0$.

THEOREM 14. *If a modular lattice L has a composition series, then every* *non-trivial element $a \in L$ is representable in the form*

$$a = a_1 \dotplus \cdots \dotplus a_m,$$

where $a_1, \ldots, a_m$ are indecomposable elements.

Proof. It is easy to deduce from Theorem 10 that a sublattice of $[0, a]$ also has a composition series. Taking note of Theorem 11, we denote by l the length of this series. For $l \leqslant 1$, the theorem is trivial. However, if $l > 1$, then we put $a = b_0$ and assume we have constructed the series

$$b_0 > \cdots > b_k,$$

where $a = b_k \dotplus c_k$ for some $c_k \in L$. If the element b_k is decomposable, then $b_k = b_{k+1} \dotplus d$, where $b_{k+1} < b_k$ and, by Theorem 7, we have

$$a = b_{k+1} \dotplus (d \dotplus c_k).$$

In view of Theorem 10, the construction described cannot be continued indefinitely. Consequently, a can be represented in the form $a = a_1 \dotplus b$, where a_1 is an indecomposable element. Clearly, the interval $[0, b]$ has composition series of length less than l. Hence,

$$b = a_2 \dotplus \cdots \dotplus a_m,$$

and, by Theorem 7,

$$a = a_1 \dotplus a_2 \dotplus \cdots \dotplus a_m.$$

Theorem 14 does not admit a converse. As an example we can take any infinite chain.

THEOREM 15. (Ore). *Let L be any modular lattice with a composition series of length l, and let*

$$1 = a_1 \dotplus \cdots \dotplus a_m = b_1 \dotplus \cdots \dotplus b_n,$$

where $a_1, \ldots, a_m, b_1, \ldots, b_n$ are indecomposable elements. Then $m = n$ and for every a_i there is an element b_j such that

$$1 = a_1 \dotplus \cdots \dotplus a_{i-1} \dotplus b_j \dotplus a_{i+1} \dotplus \cdots \dotplus a_m.$$

Proof. We proceed by induction on l. If $l = 0$, then L is a one-element lattice and the theorem holds trivially. It also holds for $l = 1$, since in this case $m = n = 1$. If $l > 1$, we put

$$\bar{a}_i = a_1 + \cdots + a_{i-1} + a_{i+1} + \cdots + a_m,$$

and consider the following three cases:

(i) $a_1 + \bar{b}_1 = \bar{a}_1 + b_1 = 1$ for suitably numbered elements b_j;

(ii) $a_1 + \bar{b}_1 < 1$ for suitably numbered elements b_j;

(iii) for all j, $a_1 + \bar{b}_j = 1$ and $\bar{a}_1 + b_j < 1$, hold.

In addition, using Theorems 10 and 13, we fix a dimension function d on L such that $d(0) = 0$.

In case (i) we have

$$d(a_1) = d(1) - d(\bar{b}_1) + d(a_1\bar{b}_1) = d(b_1) + d(a_1\bar{b}_1) \geqslant d(b_1),$$

and

$$d(b_1) = d(1) - d(\bar{a}_1) + d(\bar{a}_1 b_1) = d(a_1) + d(\bar{a}_1 b_1) \geqslant d(a_1).$$

Therefore, $d(a_1) = d(b_1)$ and it follows that $d(a_1\bar{b}_1) = d(\bar{a}_1 b_1) = 0$; hence, in view of Theorem 12, $a_1\bar{b}_1 = \bar{a}_1 b_1 = 0$. Thus, $1 = \bar{b}_1 \dot{+} a_1$, that is, a_1 can be replaced by some b_j.

In case (ii) put $q_h = (a_1 + \bar{b}_h) b_h$ $(h = 1, \ldots, n)$ and $c = q_1 + \cdots + q_n$. Since $q_h \leqslant b_h$, hence

$$c = q_1 \dot{+} \cdots \dot{+} q_n.$$

If $(a_1 + \bar{b}_1) b_1 = b_1$, then $a_1 + \bar{b}_1 \geqslant b_1$, whence

$$a_1 + \bar{b}_1 \geqslant b_1 + \bar{b}_1 = 1,$$

which is a contradiction. Consequently, $q_1 = (a_1 + \bar{b}_1) b_1 < b_1$. Therefore,

$$d(c) = d(q_1) + \cdots + d(q_n) < d(b_1) + \cdots + d(b_n) = d(1),$$

whence $c < 1$. Put $r_h = a_1 + \overline{b}_h$ and obtain $q_1 = r_1 b_1$. If

$$q_1 + \cdots + q_{i-1} = r_1 \ldots r_{i-1} (b_1 + \cdots + b_{i-1}),$$

then, taking note of the inequalities

$$b_1 + \cdots + b_{i-1} \leqslant \overline{b}_i \leqslant r_i,$$

and

$$b_i \leqslant \overline{b}_1 \ldots \overline{b}_{i-1} \leqslant r_1 \ldots r_{i-1},$$

as well as the modular law, we obtain

$$\begin{aligned} q_1 + \cdots + q_i &= r_1 \ldots r_{i-1} (b_1 + \cdots + b_{i-1}) + r_i b_i \\ &= r_i [r_1 \ldots r_{i-1} (b_1 + \cdots + b_{i-1}) + b_i] \\ &= r_1 \ldots r_i (b_1 + \cdots + b_i). \end{aligned}$$

For $i = n$ the relation proved becomes

$$c = q_1 + \cdots + q_n = r_1 \ldots r_n \cdot 1 = r_1 \ldots r_n \geqslant a_1,$$

whence

$$a_1 + \overline{a}_1 c = (a_1 + \overline{a}_1)\, c = c,$$

and, in addition,

$$a_1 \cdot \overline{a}_1 c = 0.$$

Hence

$$c = a_1 \dot{+} \overline{a}_1 c = q_1 \dot{+} \cdots \dot{+} q_n.$$

In view of Theorem 14, it can be assumed that both $\overline{a}_1 c$ and all q_h are representable as a direct sum of indecomposable elements. On the other hand, the length of a composition series of $[0, c]$ is less than l. Hence, the induction hypothesis allows us to write

$$c = e \dot{+} \overline{a}_1 c,$$

where $e \dotplus f = q_h$ for some index h. Hence,

$$e + \bar{a}_1 = e + \bar{a}_1 c + \bar{a}_1 = c + \bar{a}_1 = a_1 + \bar{a}_1 c + \bar{a}_1 = 1,$$

and, consequently,

$$0 \neq e \leqslant q_h \leqslant b_h.$$

However,

$$e + b_h \bar{a}_1 = b_h (e + \bar{a}_1) = b_h,$$

and

$$e \cdot b_h \bar{a}_1 = e c b_h \bar{a}_1 = [e (\bar{a}_1 c)]\, b_h = 0,$$

that is,

$$b_h = e \dotplus b_h \bar{a}_1.$$

Since b_h is indecomposable, it follows that $e = b_h$. Thus

$$1 = b_h + \bar{a}_1,$$

that is, in case (ii) a_1 can also be replaced by some b_h. Moreover, it is found that $h \neq 1$. In fact, since $q_1 < b_1$, b_1 cannot be a direct summand in q_1, and for $h \neq 1$ we have $b_1 q_h \leqslant b_1 b_h = 0$.

In case (iii) we have, in particular,

$$b_1 + \bar{a}_1 < 1.$$

From the conclusion obtained while considering case (ii), it therefore follows that

$$1 = a_h \dotplus \bar{b}_1,$$

and so $h \neq 1$. However,

$$1 = a_h \dotplus \bar{a}_h.$$

Hence, there exists a perspective mapping

$$\chi = P_{(\bar{b}_1 \to \bar{a}_h;\ a_h)}.$$

Consequently,

$$\bar{a}_h = a_1 \dot{+} \cdots \dot{+} a_{h-1} \dot{+} a_{h+1} \dot{+} \cdots \dot{+} a_m = \chi(b_2) \dot{+} \cdots \dot{+} \chi(b_n),$$

where the terms on the right-hand side are indecomposable. Applying the result to the lattice $[0, \bar{a}_h]$ we obtain

$$\bar{a}_h = \chi(b_k) \dot{+} a_2 \dot{+} \cdots \dot{+} a_{h-1} \dot{+} a_{h+1} \dot{+} \cdots \dot{+} a_m$$

for some index k. Hence

$$\chi(b_k) + a_h = (b_k + a_h)\,\bar{a}_h + a_h = (b_k + a_h)(\bar{a}_h + a_h) = b_k + a_h$$

and, consequently

$$\begin{aligned} b_k + \bar{a}_1 &= b_k + a_h + a_2 + \cdots + a_{h-1} + a_{h+1} + \cdots + a_m \\ &= a_h + \chi(b_k) + a_2 + \cdots + a_{h-1} + a_{h+1} + \cdots + a_m \\ &= a_h + \bar{a}_h = 1. \end{aligned}$$

We obtain a contradiction to the hypothesis, showing that case (iii) is impossible.

Finally, assume that $m < n$. Then by applying what has been proved above, we successively replace the elements $a_1, a_2, \ldots, a_m$ by some b_j. As a result, by changing, if necessary, the numbering of the b_j, we arrive at the equality

$$b_1 + \cdots + b_p = b_1 + \cdots + b_n,$$

where $p \leqslant m < n$. Hence

$$b_n = (b_1 + \cdots + b_p)\, b_n = 0,$$

contradicting the indecomposability of the element b_n.

Ore's Theorem when applied to a lattice of subspaces of a finite-dimensional linear space yields the well known exchange theorem for two equivalent systems of linearly independent vectors. From this it also follows that all bases contain precisely the same number of vectors. A useful consequence of Ore's theorem is the application to the lattices of left and two-sided ideals of an Artin's ring with a unit element.

An element a of the lattice L is called $\cap$-*irreducible*, if $a = bc$ implies either $a = b$ or $a = c$. A representation of the element a in the form $a = a_1 \ldots a_n$ is called an *irredundant* $\cap$-*representation*, if the elements $a_1, \ldots, a_m$ are $\cap$-irreducible, and

$$a_1 \ldots a_{i-1}a_{i+1} \ldots a_n \nleqslant a_i$$

for all $i = 1, \ldots, n$.

THEOREM 16. (Kurosh–Ore). *If*

$$w = a_1 \ldots a_m = b_1 \ldots b_n$$

are two irredundant $\cap$-*representations of an element* w *of a modular lattice* L, *then* $m = n$ *and for every* a_i *there is an element* b_j, *such that*

$$w = a_1 \ldots a_{i-1}b_ja_{i+1} \ldots a_m.$$

Proof. Put

$$\bar{a}_i = a_1 \ldots a_{i-1}a_{i+1} \ldots a_m,$$

and

$$c_j = \bar{a}_1 b_j \qquad (j = 1, \ldots, n).$$

Then

$$w \leqslant c_1 \ldots c_n \leqslant b_1 \ldots b_n \leqslant w,$$

that is,

$$w = c_1 \ldots c_n.$$

From Theorem 9, it follows that $[a_1\bar{a}_1, \bar{a}_1]$ and $[a_1, a_1 + \bar{a}_1]$ are isomorphic intervals. The element a_1 is $\cap$-irreducible as an element of the lattice $[a_1, a_1 + \bar{a}_1]$. However, then, the element $w = a_1\bar{a}_1 = c_1 \ldots c_n$ is $\cap$-irreducible in the lattice $[a\bar{a}_1, \bar{a}_1]$. Since, $c_j \leqslant \bar{a}_1$ for all j, it follows that $w = c_j$ for some index j, that is,

$$w = b_ja_2 \ldots a_m.$$

Assuming that $m < n$ and, using the conclusion just obtained, we successively replace the elements $a_1, \ldots, a_m$ by some b_j. As a result, by renumbering, if necessary, the b_j, we arrive at the equality

$$b_1 \ldots b_p = b_1 \ldots b_n,$$

where $p \leqslant m < n$, contradicting the fact that the representation

$$w = b_1 \ldots b_n$$

is $\cap$-irreducible.

The Kurosh–Ore theorem also finds fruitful applications in the ideal theory of rings.

An element p of a lattice L with zero is called an *atom* if p is a minimal element in the set of non-trivial elements. The lattice L with zero is called *atomic*, if for every $0 \neq a \in L$ there is an atom $p \leqslant a$.

THEOREM 17. *Let L be a modular lattice and let $a = b + p_1 + \ldots + p_n$, p_i being atoms. Then for suitable numbering we have*

$$a = b \dotplus p_1 \dotplus \cdots \dotplus p_m$$

($m = 0$ is possible).

Proof. Without loss of generality it can be assumed that

$$p_i \nleqslant b + p_1 + \cdots + p_{i-1} + p_{i+1} + \cdots + p_n$$

for all i. If $b + p_1 + \ldots + p_n$ is not a direct sum, then by Theorem 5, we have

$$0 < (b + p_1 + \cdots + p_{k-1})\, p_k \leqslant p_k$$

for some k. Since p_k is an atom, it follows that $(b+p_1 + \cdots + p_{k-1})\, p_k = p_k$ and, consequently

$$p_k \leqslant (b + p_1 + \cdots + p_{k-1}) + (p_{k+1} + \cdots + p_n),$$

which contradicts our assumption.

THEOREM 18. *If*

$$a = p_1 + \cdots + p_n,$$

where $p_1, \ldots, p_n$ are atoms of a modular lattice, then the lattice $[0, a]$ has a composition series.

Proof. By Theorem 17

$$a = p_1 \dot{+} \cdots \dot{+} p_m,$$

so we can prove the theorem in hand by induction on m. From $m = 1$, the theorem is trivial, since $[0, p_1]$ is a simple interval. Now, assume that the lattice $[0, p_1 + \cdots + p_{m-1}]$ has a composition series. From Theorem 9, it follows that $[p_1 + \cdots + p_{m-1}, a]$ and $[p_m (p_1 + \cdots + p_{m-1}), p_m] = [0, p_m]$ are isomorphic intervals. Since the latter interval is simple, the proof is complete.

THEOREM 19. *For a modular lattice L, having a composition series, the following properties are equivalent*:

(1) *L is a complemented lattice*;

(2) *every element in L is representable as a direct sum of atoms*;

(3) 1 *is representable as a sum of atoms.*

Proof. If L is a complemented lattice, a is an indecomposable element in L and if $0 \leqslant b < a$, then, by Theorem 3, we have $a = b \dot{+} c$, whence $a = c$ and, consequently, $b = ab = bc = 0$. Thus, element a is found to be an atom, and property (2) follows from Theorem 14. The implication (2) ⇒ (3) is obvious. Finally, let

$$1 = p_1 + \cdots + p_n,$$

where the p_i are atoms and let $a \in L$. Then

$$1 = a + p_1 + \cdots + p_n,$$

and, in view of Theorem 17,

$$1 = a \dot{+} (p_1 + \cdots + p_m),$$

proving the validity of property (1).

We shall now prove a theorem similar to Theorem 19 for complete lattices. As a preliminary we define a *directed set* as a poset Ω, in which for any $\alpha, \beta \in \Omega$ there is an element $\gamma \in \Omega$, such that $\gamma \geqslant \alpha, \beta$. A complete lattice L is called *continuous*, if

$$a \sup \Omega = \sup (a\Omega)$$

for every $a \in L$ and for any directed set $\Omega \subseteq L$. It is easily verified that a complete lattice, having a composition series, is continuous†.

THEOREM 20. *Let L be a complete modular lattice. Then*:

(*a*) *if L is atomic and complemented, then every nonzero element in L is representable as a* LUB *of some set of atoms;*

(*b*) *if L is continuous and its unit element is a* LUB *of the set of all atoms, then L is atomic and complemented.*

Proof. (*a*) Let $0 \neq a \in L$, $A = \{p \mid p \text{ an atom}, p \leqslant a\}$ and let $b = \sup A$. Clearly, $a \geqslant b$. If $a \neq b$, then by Theorem 3, $a = b \dotplus c$, where $c \neq 0$. By the hypothesis there is an atom q such that $q \leqslant \leqslant c \leqslant a$. However, then $q \in A$, whence $q \leqslant bc = 0$. The contradiction obtained shows that $a = \sup A$.

(*b*) Let A be the set of all atoms, and Ω the set of all finite sums of atoms. If $0 \neq a \in L$, then, since Ω is a directed set, we have

$$a = a1 = a \sup A = a \sup \Omega = \sup (a\Omega).$$

Consequently, $as \neq 0$ for some $s \in \Omega$. Let

$$s = p_1 + \cdots + p_m,$$

where p_i are atoms. From Theorems 18 and 19 follows the existence of an atom q, such that $q \leqslant as \leqslant a$, showing that L is an atomic lattice. Again, let $a \in L$. Consider the set $B = \{x \mid ax = 0\}$. If C is a chain in B, then since L is a continuous lattice, we have

$$a \cdot \sup C = \sup aC = \sup 0 = 0,$$

†Strictly speaking, in place of 'directed' and 'continuous', one should say 'upward directed' and 'upper continuous'. This is not done because the dual concept is not considered in the usual text.

that is, sup $C \in B$. Hence, the Kuratowski–Zorn lemma enables us to find in B a maximal element b. Of course, $ab = 0$. If $a + b \neq 1$, then $p \nleq a + b$, for some atom p. However, then $p(a + b) < p$ and, consequently, $p(a + b) = 0$. In view of Theorem 5, it follows that $a(b + p) = 0$. The relation so obtained contradicts the maximality of the element b, completing the proof.

A modular lattice F is called the *free modular lattice* with a system of free generators P, if:

(i) $P \subseteq F$;

(ii) F is generated by the set P;

(iii) for any mapping φ of the set P into an arbitrary modular lattice L, there is a homomorphism ψ of the lattice F into the lattice L, such that $\psi(x) = \varphi(x)$ for all $x \in P$.

The free modular lattice, whose system of free generators consists of n elements, is denoted by $FML(n)$.

THEOREM 21. *Any non-void set P can serve as a system of free generators for some free modular lattice*†.

Proof. Consider the set $\mathfrak{M}$ of all modular lattices, generated by a subset of power not exceeding the power of the set P. Here each lattice is taken as often as there exist mappings of P onto the corresponding system of generators. In addition, consider the direct product of the lattices under consideration. It is easily verified that this is a modular lattice. Let us denote it by G. Further, to every element x in P we let correspond the series $\Phi(x) \in G$ composed of its images under the mappings indicated above, and we denote by F the sublattice of G, generated by the set $\Phi(P)$. Since the power of the direct product P of copies of a two-element chain is not less than the power of the set P, among the lattices of the set $\mathfrak{M}$ there exists one such that the mapping of P onto its system of generators is one-to-one. It follows that the mapping Φ is also one-to-one. Hence, the set P can be identified

†Note that Theorem 21 is a particular case of a similar theorem in the theory of universal algebra.

with $\Phi(P)$. Thus it becomes clear that the modular lattice F has properties (i) and (ii) in the definition of the free modular lattice. If φ is a mapping of the set P into a modular lattice L, then a sublattice of L, generated by the set $\varphi(P)$, together with the mapping φ exists in the set $\mathfrak{M}$. Let π be the projection of the lattice G on the corresponding direct factor. For any $x \in P$ we have $\varphi(x) = \pi(\Phi(x))$, that is, by putting $\psi = \pi\Phi$ the validity of property (iii) is proved.

THEOREM 22. *The lattice FML (3) is finite.*

Proof. Let x, y and z be the free generators of the lattice $F = FML\,(3)$ and of the lattice $G = FL\,(3)$. From property (v) of the definition of a free extension, it is easy to deduce that there exists a homomorphism φ of lattice G onto the lattice F. Taking note of the method of constructing the lattice G described in Chap. 5, we consider the following 28 words: $x, y, z, x + y, x + z, y + z, x + y + z, xy, xz, yz, xyz, (x + y)(x + z), (x + y)(y + z), (x + z)(y + z), xy + xz, xy + yz, xz + yz, x + yz, y + xz, z + xy, x(y + z), y(x+z), z(x + y), (x + y)(x + z)(y + z), xy + xz + yz, (x + yz)(y + z), (y + xz)(x + z), (z + xy)(x + y)$, which are identified with their images under the homomorphism φ. Lengthy, though not very tedious calculatious show that the addition and multiplication in the lattice F do not go beyond the limit of this 28-element set, proving the theorem. Of course, for this enumeration it has to be kept in view that the lattice F obeys the modular law.

REMARK. The above reasoning establishes that the lattice FML (3) contains not more than 28 elements. In fact, it can be shown that it contains precisely 28 elements.

THEOREM 23. *The lattic FML (4) is infinite.*

Proof. It is easy to see that every modular lattice with four generators is a homomorphic image of the lattice FML (4). The theorem will be proved thereby if we are able to indicate even a single infinite modular lattice with four generators. To this end, from the lattice of all subspaces of a finite-dimensional linear space over some field with the basis $e_1, e_2, \ldots$ we select a sublattice L, generated by the following four subspaces:

S_1 — the linear hull of the vectors e_{2k}, $k = 1, 2, \ldots$;

S_2 — the linear hull of the vectors $e_{2k} + e_{2k+1}$, $k = 1, 2, \ldots$;

S_3 — the linear hull of the vectors e_{2k-1}, $k = 1, 2, \ldots$;

S_4 — the linear hull of the vectors $e_{2k-1} + e_{2k}$, $k = 1, 2, \ldots$.

Let T_r be the linear hull of the vectors $e_{2r}, e_{2r+2}, \ldots$. Simple calculations show that

$$T_{r+1} = [(T_r + S_2) \cap S_3 + S_4] \cap S_1.$$

Since $T_1 = S_1 \in L$, it is easy to deduce that $T_r \in L$ for $r = 1, 2, \ldots$. Since all subspaces of T_r are distinct, it is proved that the lattice L is infinite.

Exercises

1. Prove that the modularity of the lattice L is equivalent to each of the following properties:

   (*a*) $a + b(a + c) = (a + b)(a + c)$; (*b*) $(a + bc)(b + c) = a(b + c) + bc$; (*c*) if $a \leqslant c$, and $d \leqslant b$, then $a + b(c + d) = (a + b)c + d$; (*d*) if $a \leqslant c \leqslant a + b$, then $a + bc = c$; (*e*) if $a \leqslant b \leqslant c + d$, $ac = bc$ and $(a+c)d = (b + c)d$, then $a = b$; (*f*) L does not contain a sublattice depicted in Fig. 9.

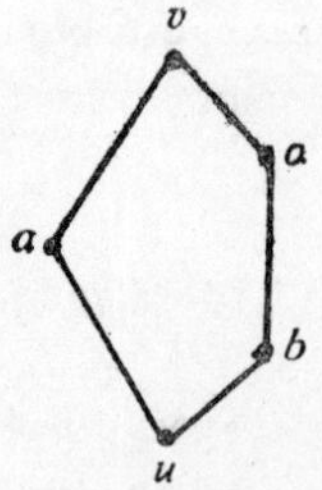

Figure 9.

2. If x and y are the complements af an element z in the interval $[a, b]$ of a modular lattice and if $x \leqslant y$ then $x = y$.
3. In every modular lattice, the equality $(ab + ac)(ab + bc) = ab$ holds, and the relation $(a + b)c = bc$ implies that $a(b + c) = ab$.

4. In a modular lattice with zero, the equality $(a_1 + \cdots + a_n)\, b = 0$ implies $(a_1 + b) \ldots (a_n + b) = a_1 a_2 \ldots a_n + b$.

5. If $a_1, \ldots, a_n, b_1, \ldots, b_n$ are elements of a modular lattice and if $a_i \leqslant b_j$ for $i \neq j$ then

$$(a_1 + \cdots + a_n)\, b_1 \ldots b_n = a_1 b_1 + \cdots + a_n b_n.$$

6. The ordinal sum of modular lattices is also modular.

7. The modularity of a lattice is equivalent to the modularity of the lattice of its ideals.

8. The elements a and b of a modular lattice with 0 and 1 have complements, if complements exist for the elements ab and $a + b$.

9. If C is a chain in a complemented modular lattice L, having a composition series, then L contains a chain C^*, dual to C, and consisting of the complements of its elements.

10. A complemented modular lattice, satisfying the maximum condition, also satisfies the minimum condition.

11. The relation

$$a \equiv b \underset{\text{def}}{=} \boxed{\begin{array}{l}\text{the interval } [ab,\ a + b] \\ \text{contains a finite maximal} \\ \text{chain}\end{array}}$$

on a modular lattice is a congruence relation.

12. If d and d' are dimension functions on a modular lattice, then $d - d' =$ constant.

13. Formulate and prove the dual of the Kurosh–Ore theorem.

14. Study the lattice of Fig. 10 and verify that one cannot omit modularity from the hypothesis of the Kurosh–Ore theorem.

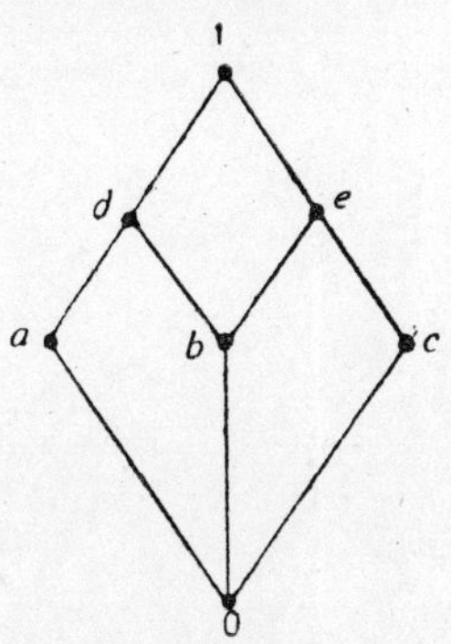

Figure 10.

15. Let L be a modular lattice, $a \in L$ and let $a = p_1 \dotplus \cdots \dotplus p_m = q_1 \dotplus \ldots \dotplus q_n$, where p_i, q_j are atoms. Prove that $m = n$ and $a = p_1 \dotplus p_{m-1} \dotplus q_j$ for some index j.

16. Any interval of an atomic complemented modular lattice is an atomic lattice.

17. Prove the uniqueness of the free modular lattice with a given system of free generators.

18. A sublattice of a free modular lattice, generated by some of the free generators, is free.

19. For any family $\{L_\alpha \mid \alpha \in \Omega\}$ of modular lattices, there exists a modular lattice L, generated by the union $\bigcup_{\alpha \in \Omega} L_\alpha$ such that, for any set $\{\varphi_\alpha \mid \alpha \in \Omega\}$ of the homomorphisms of the lattices L_α into any modular lattice M there exists a homomorphism of L into M, coinciding with φ_α on each of the L_α.

7

DISTRIBUTIVE LATTICES

A lattice L is called *distributive* if for any $a, b, c \in L$, there holds

$$(a + b)\, c = ac + bc.$$

An important example of a distributive lattice is the lattice of all subsets of an arbitrary set. However, the modular lattice of all subspaces of a linear space is not distributive. Any chain is also a distributive lattice.

THEOREM 1. *The following properties of the lattice L are equivalent*:

(i) *L is distributive*;

(ii) $ab + c = (a + c)(b + c)$ *for any* $a, b, c \in L$;

(iii) $ab + bc + ca = (a + b)(b + c)(c + a)$ *for any* $a, b, c \in L$;

(iv) *if for some* $c \in L$, $a + c = b + c$ *and* $ac = bc$ *hold, then* $a = b$.

Proof. (i) ⇒ (ii). We have

$$(a + c)(b + c) = (a + c)\, b + c = ab + bc + c = ab + c.$$

(ii) ⇒ (iii). Indeed,

$$\begin{aligned} ab + bc + ca &= (a + bc + ca)(b + bc + ca) = (a + bc)(b + ca) \\ &= (b + a)(c + a)(b + c)(b + a) \\ &= (a + b)(b + c)(c + a). \end{aligned}$$

(iii) ⇒ (i). If $a \leqslant c$, then

$$ac + bc = ab + bc + ca = (a + b)(b + c)(c + a)$$
$$= c(a + b)(b + c) = c(a + b).$$

Further, we put $u = ab + bc + ac$ and $v = (a + b)(b + c)(a + c)$. By hypothesis, $u = v$ and, consequently, $cu = cv$.

Taking note of all that has been said above, we obtain

$$cu = c(ab + bc + ac) = c(bc + ac) + abc = ac + bc,$$

and

$$cv = c(a + b)(b + c)(a + c) = c(a + b)$$

which was to be proved.

(i) ⇒ (iv). If $a + c = b + c$ and $ac = bc$, then, applying (i) we get

$$a = a(a + c) = a(b + c) = ab + ac = ab + bc$$
$$= (a + c)b = (b + c)b = b.$$

(iv) ⇒ (iii). Put $u = ab + bc + ac$, $v = (a + b)(b + c)(a + c)$, $p = ac + b(a + c)$, $q = bc + a(b + c)$ and $r = ab + c(a + b)$. By Theorem 6.1, property (iv) implies that the modular law holds. Hence,

$$p + r = [ac + b(a + c)] + [ab + c(a + b)]$$
$$= b(a + c) + c(a + b) = (a + b)[b(a + c) + c]$$
$$= (a + b)(a + c)(b + c) = v,$$

$$q + r = [bc + a(b + c)] + [ab + c(a + b)]$$
$$= a(b + c) + c(a + b) = (a + b)[a(b + c) + c]$$
$$= (a + b)(b + c)(a + c) = v,$$

$$pr = [ac + b(a + c)][ab + c(a + b)]$$
$$= ab + [ac + b(a + c)]c(a + b)$$
$$= ab + [ac + b(a + c)c](a + b) = u,$$

and

$$qr = [bc + a(b + c)][ab + c(a + b)] = ab + [bc + a(b + c)]c(a + b)$$
$$= ab + [bc + a(b + c)c](a + b) = u.$$

In view of (iv) we have $p = q$. But, then

$$p = p + q = [ac + b(a + c)] + [bc + a(b + c)]$$
$$= b(a + c) + a(b + c) = (a + c)[b + a(b + c)] = v,$$

and

$$p = pq = [ac + b(a + c)][bc + a(b + c)]$$
$$= bc + [ac + b(a + c)]a(b + c)$$
$$= bc + [ac + b(a + c)a](b + c) = u.$$

Thus, we have $u = p = v$, as we wished to show.

From property (iii) of Theorem 1 it is seen that a poset, which is the dual of a distributive lattice, is itself a distributive lattice. Hence, the duality principle applies to a distributive lattice.

THEOREM 2. *Every ideal S and every filter F of a distributive lattice L is standard.*

Proof. If $a, b \in L$, $s \in S$ and $a \leqslant b + s$, then $a = a(b + s) = ab + as$. Clearly, $ab \leqslant b$ and $as \in S$. The second assertion holds by duality.

In the case of distributive lattices, one is able to solve completely the problem of the relationship between the ideals and the homomorphisms.

THEOREM 3. (Hashimoto). *The following properties of a lattice L are equivalent:*

(i) *L is distributive;*

(ii) *every ideal of the lattice L is the kernel of some congruence;*

(iii) *every filter of L is the kernel of some congruence;*

(iv) *every convex sublattice of L is a residue class of some congruence.*

Proof. The implications (i) $\Rightarrow$ (ii) and (i) $\Rightarrow$ (iii) are immediate from Theorems 2, 4.9 and 4.11.

If properties (ii) and (iii) hold, then so does property (iv) by Theorem 4.8, proving the implication (i) $\Rightarrow$ (iv). The implication (iv) $\Rightarrow$ (ii) is trivial. It remains to establish the validity of (ii) $\Rightarrow$ (i) and (iii) $\Rightarrow$ (i). For this it suffices to prove the former, whence the latter will follow by duality. For the proof, assume that the lattice L has property (ii), but is not modular. Then, we can find elements $a, b, c \in L$ such that $a \leqslant c$ and $(a + b) c > a + bc$. By hypothesis, the principal ideal $(a + bc)^{\triangledown}$ is the kernel of some congruence $\equiv$. However, since, $a \in (a + bc)^{\triangledown}$, we get

$$(a + b) c \equiv bc \equiv a + bc.$$

Hence,

$$(a + b) c \in (a + bc)^{\triangledown},$$

contradicting the assumption. Thus, L is a modular lattice. If it is not distributive, then for some $a, b, c \in L$, we have

$$(a + b) c > ac + bc.$$

If $\equiv$ is a congruence relation with the kernel $(a + bc)^{\triangledown}$, then

$$(a+ b) c = (a + bc + b) c \equiv bc,$$

whence $(a + b) c \leqslant a + bc$. Hence, keeping in view the modular law, we obtain

$$(a + b) c \leqslant (a + bc) c = ac + bc,$$

violating the assumption.

THEOREM 4. (Kolibiar–Hashimoto–Grätzer–Schmidt). *The following properties of a distributive lattice L are equivalent*:

(i) *L is a complemented lattice*;

(ii) *every ideal of L is the kernel of at least one congruence relation on L*;

(iii) *every filter of L is the kernel of at least one congruence relation on L*;

(iv) *every convex sublattice of L is a residue class of at least one congruence relation on L;*

(v) *any two congruence relations on L are permutable.*

Proof. The implication (i) $\Rightarrow$ (iv) is immediate from the preceding Theorem 3 and Theorem 4.6. By Theorem 4.7, the implication (i) $\Rightarrow$ (v) holds. The implications (iv) $\Rightarrow$ (ii) and (iv) $\Rightarrow$ (iii) are trivial. Further, assume that the property (v) is satisfied and consider an element c in the interval $[a, b]$. By Theorem 2, ideal c^∇ and filter c^Δ are respectively, the kernel of the congruences $\equiv$ and $\approxeq$, described in Theorems 4.9 and 4.11. Since $a \equiv c$ and $c \approxeq b$, hence, by property (v), for some element d, we have $a \approxeq d$ and $d \equiv b$. These relations show the validity of the equalities $ad = (a + d)\, v$ and $d + b = db + u$, where $u \leqslant c \leqslant v$. Hence,

$$a = ac \leqslant av \leqslant ad \leqslant d \leqslant d + b \leqslant b + u \leqslant b + c = b,$$

and, consequently,

$$c + d = c + d + db + u = c + d + b = b,$$

and

$$cd = cd\,(a + d)\, v = cda = a.$$

Thus, d is a complement of c in the interval $[a, b]$, that is, the implication (v) $\Rightarrow$ (i) is established. It remains to demonstrate the validity of the implications (ii) $\Rightarrow$ (i) and (iii) $\Rightarrow$ (i). We shall confine ourselves to the former of these, as the latter follows by duality. To this end, we take in the interval $[a, b]$ an element c different from a and b. Let S be the set of all s in L, such that $cs \leqslant a$. Since L is a distributive lattice, it is easy to verify that S is an ideal. By virtue of Theorem 2, this ideal is the kernel of the congruence θ, described in Theorem 4.9.

From Theorem 2 of the present chapter and Theorem 4.11 it follows that on the quotient lattice $\bar{L} = L/\theta$, there exists a congruence relation $\bar{\theta}$ with kernel filter $\bar{c}^\Delta$. Denote by ρ the kernel of the homomorphism of L onto the quotient lattice $\bar{L}/\bar{\theta}$, and by T the kernel ideal of ρ. Of course, $S \subseteq T$. If $t \in T$ then, by the definition of $\bar{\theta}$, we obtain

$$\bar{0} = \bar{t}\,\bar{0} = (\bar{t} + \bar{0})\,\bar{u} = \bar{t}\,\bar{u},$$

where $\overline{cu} = \overline{c}$, and $\overline{0}$ is the zero of $\overline{L}$. Therefore $\overline{0} = \overline{tc}$ and, consequently, $tc\theta s$, where $s \in S$. The latter implies that

$$tc \leqslant tc + s = tcs + s' \leqslant ta + s' \leqslant a + s',$$

where $s' \in S$. Multiplying by c, we obtain

$$tc \leqslant ac + s'c \leqslant a,$$

which means that $t \in S$. By property(ii), the equality $\theta = \rho$ follows from the equality $S = T$. Consequently, $\overline{b}\,\overline{\theta}\,\overline{c}$ implies $b\rho c$ and, hence, also $b\theta c$. Thus,

$$b = b + c = bc + s = c + s,$$

where $s \in S$. However, then,

$$c + (a + s) = c + s = b,$$

and

$$a \leqslant c\,(a + s) = ca + cs \leqslant a,$$

that is,

$$c\,(a + s) = a.$$

Thus, $a + s$ forms a complement of element c in the interval $[a, b]$, completing the proof.

THEOREM 5. (Stone). *If a and b are two distinct elements of a distributive lattice L, then there exists a congruence relation θ, such that $a \not\theta\, b$ and the quotient lattice L/θ is a two-element chain.*

Proof. Without the loss of generality, it can be assumed that $b \nleqslant a$. Hence the set $\mathfrak{P}$ of ideals I of L such that $a \in I$ and $b \overline{\in} I$, contains the ideal $a^{\triangledown}$; the set $\mathfrak{P}$ has a natural partial order. If

$$\mathfrak{C} : I_1 \subseteq \ldots \subseteq I_\alpha \subseteq \ldots$$

is a chain in $\mathfrak{P}$, then it is easy to verify the union $\cup\, I_\alpha \in \mathfrak{C}^{\Delta}$. By the Kuratowski–Zorn lemma, $\mathfrak{P}$ contains a maximal element H. By Theorem 3, the ideal H is the kernel of a congruence $\equiv$. Of course, $a \not\equiv b$. If $x \overline{\in} H$, then it is easy to deduce that the set J of all sums of the form

$y + s$, where $y \leqslant x$, $s \in H$, is an ideal. Since $H \underset{\neq}{\subset} J$, hence, $b \in J$. Therefore, $b = y + s$, where $y \leqslant x$, $s \in H$ and, consequently,

$$b \equiv y \leqslant x.$$

Thus the nonzero elements of the quotient lattice $L/\equiv$ form a filter which is, obviously, the kernel of the homomorphism of the lattice $L/\equiv$ into a two-element chain C. The kernel of the homomorphism L onto C is the desired congruence relation,

THEOREM 6. (Birkhoff). *Every distributive lattice L is isomorphic with a lattice of subsets of some set* (not necessarily all of them).

Proof. Consider the set $\mathfrak{M}$ of all possible pairs (a, b), where a and b are distinct elements of L. For every pair (a, b), by using Theorem 5, we can find a congruence relation $\theta_{a,b}$ such that $a \not\theta_{a,b}\, b$ and the quotient lattice $L/\theta_{a,b}$ is a two-element chain. Denote by $\varphi_{a,b}$ the natural homomorphisms of L onto $L/\theta_{a,b}$. If $x \in L$, then put

$$\Phi(x) = \{(a, b) \mid (a, b) \in \mathfrak{M},\ \varphi_{a,b}(x) = 1\}.$$

Thus, every element of the lattice L corresponds to a subset of the set $\mathfrak{M}$. Let $x, y \in L$. If $x \neq y$, then $\varphi_{x,y}(x) \neq \varphi_{x,y}(y)$ and, consequently, $\Phi(x) \neq \Phi(y)$. Further, $\varphi_{a,b}(x + y) = 1$, iff either $\varphi_{a,b}(x) = 1$, or $\varphi_{a,b}(y) = 1$. Hence, $(a, b) \in \Phi(x + y)$ iff $(a, b) \in \Phi(x)$ or, $(a, b) \in \Phi(y)$. Consequently, $\Phi(x + y) = \Phi(x) \cup \Phi(y)$. Similarly, $\varphi_{a,b}(xy) = 1$ iff $\varphi_{a,b}(x) = \varphi_{a,b}(y) = 1$, whence $\Phi(xy) = \Phi(x) \cap \Phi(y)$. Thus, Φ forms the desired lattice isomorphism.

By a simple induction we can show that in every distributive lattice the following identities hold:

$$a \sum_{i=1}^{m} b_i = \sum_{i=1}^{m} ab_i,$$

$$a + \prod_{i=1}^{m} b_i = \prod_{i=1}^{m} (a + b_i),$$

$$\sum_{i=1}^{m} a_i \sum_{j=1}^{n} b_j = \sum_{i=1}^{m} \sum_{j=1}^{n} a_i b_j$$

and

$$\prod_{i=1}^{m} a_i + \prod_{j=1}^{n} b_j = \prod_{i=1}^{m}\prod_{j=1}^{n} (a_i + b_j).$$

The following relations are a little harder to prove:

$$\prod_{i=1}^{m}\sum_{j=1}^{n_i} b_{ij} = \sum_{\varphi\in\Phi}\prod_{i=1}^{m} b_{i\varphi(i)} \quad\text{and}\quad \sum_{i=1}^{m}\prod_{j=1}^{n_i} b_{ij} = \prod_{\varphi\in\Phi}\sum_{i=1}^{m} b_{i\varphi(i)},$$

where Φ runs through all the functions φ, associating with each i in the set $\{1, \dots, m\}$ a number $\varphi(i) \in \{1, \dots, n_i\}$. For example, let us establish the former relation. For $m = 2$ it coincides with one of the equalities written above. Furthermore, we have

$$\prod_{i=1}^{m}\sum_{j=1}^{n_i} b_{ij} = \left(\prod_{i=1}^{m-1}\sum_{j=1}^{n_i} b_{ij}\right)\sum_{j=1}^{n_m} b_{mj} = \left(\sum_{\varphi\in\Phi'}\prod_{i=1}^{m-1} b_{i\varphi(i)}\right)\sum_{j=1}^{n_m} b_{mj} =$$

$$= \sum_{j=1}^{n_m}\sum_{\varphi\in\Phi'}\left(\prod_{i=1}^{m-1} b_{i\varphi(i)}\right) b_{mj} = \sum_{\varphi\in\Phi}\prod_{i=1}^{m} b_{i\varphi(i)},$$

where the set Φ' is obtained by restricting all the functions in Φ to the set $\{1, \dots, m-1\}$. In the case of complete lattices, we can also express the *infinite distributive law* by writing infinite analogues of the identities considered earlier:

$$a \cdot \sup B = \sup (aB), \tag{7.1}$$

$$a + \inf B = \inf (a + B), \tag{7.1'}$$

$$\sup A \cdot \sup B = \sup (AB), \tag{7.2}$$

$$\inf A + \inf B = \inf (A + B), \tag{7.2'}$$

$$\inf_{\alpha\in A}\sup_{\beta\in B_\alpha} \{a_{\alpha\beta}\} = \sup_{\varphi\in\Phi}\inf_{\alpha\in A} \{a_{\alpha\varphi(\alpha)}\}, \tag{7.3}$$

$$\sup_{\alpha\in A}\inf_{\beta\in B_\alpha} \{a_{\alpha\beta}\} = \inf_{\varphi\in\Phi}\sup_{\alpha\in A} \{a_{\alpha\varphi(\alpha)}\}, \tag{7.3'}$$

where

$$AB = \{ab \mid a \in A, b \in B\}, A + B = \{a + b \mid a \in A, b \in B\},$$

Φ is the set of all single-valued functions φ, associating with $\alpha \in A$ the element $\varphi(\alpha) \in B_{\alpha}$. It can be shown that the identities (7.1) and (7.1′) (and hence, all the rest also) are not satisfied in every complete distributive lattice and that in distinction to the finite case, they do not imply each other. For proof, consider a set of non-negative integers, ordered by divisibility (that is, $a \leqslant b \underset{\text{def}}{=} a$ divides b) assuming also that 0 divides 0. It is easy to verify that $\sup A = \underset{a \in A}{\text{LCM}} \{a\}$ and $\inf A = \underset{a \in A}{\text{HCF}} \{a\}$. A quite routine check of (7.1′) shows, in particular, that the resultant lattice is distributive. However,

$$\text{HCF}\,\{2, \underset{1 \leqslant k < \infty}{\text{LCM}}\,(2k - 1)\} = \text{HCF}\,\{2, 0\} = 2,$$

and

$$\underset{1 \leqslant k < \infty}{\text{LCM}}\,\{\text{HCF}\,\{2, (2k - 1)\}\} = \underset{1 \leqslant k < \infty}{\text{LCM}}\,\{1\} = 1,$$

that is, the equality (7.1) does not hold.

THEOREM 7. *The following properties of a complete lattice I are equivalent*:

(i) *in L the identity* (7.1) *holds*;

(ii) *in L the identity* (7.2) *holds*;

(iii) *L is continuous and distributive.*

Proof. The implications (i) ⇒ (iii) and (ii) ⇒ (i) are trivial. If (7.1) holds, then

$$\sup A \cdot \sup B = \sup_{a \in A} (a \sup B) = \sup_{a \in A} \sup (aB) = \sup AB,$$

that is, (7.2) remains valid. If (iii) holds, then we consider a system $\{B_\alpha\}$ of all finite subsets of the set B, and if $B_\alpha = \{b_{\alpha 1}, \ldots, b_{\alpha n_\alpha}\}$, then we put $b_\alpha = b_{\alpha 1} + \cdots + b_{\alpha n_\alpha}$. Clearly, $\Omega = \{b_\alpha\}$ is a directed set. Hence, from (iii) and Theorem 1.4, it follows that

$$a \cdot \sup B = a \sup \Omega = \sup (a\Omega) = \sup_{\alpha} \{ab_\alpha\}$$

$$= \sup_{\alpha} \{ab_{\alpha 1} + \cdots + ab_{\alpha n_\alpha}\} = \sup aB,$$

giving the desired conclusion.

THEOREM 8. *In every complete lattice L, the equality* (7.3) *implies* (7.1), *and* (7.3′) *implies* (7.1′).

Proof. By duality considerations it will suffice to prove either of the implications. Assume that the equality (7.3) holds. If $A = \{1, 2\}$ $B_1 = \{1\}$, $B_2 = B$, $a_{11} = a$ and $a_{2b} = b$ for all $b \in B$, then for every function φ from the set Φ indicated in (7.3), we have $\varphi(1) = 1$ and $\varphi(2) = b(\varphi)$ for some $b(\varphi) \in B$. Hence

$$a \sup B = \inf_{i=1,2} \sup_{\beta \in B_i} \{a_{i\beta}\} = \sup_{\varphi \in \Phi} \inf_{i=1,2} \{a_{i\varphi(i)}\}$$

$$= \sup_{\varphi \in \Phi} a_{11} a_{2\varphi(2)} = \sup_{\varphi \in \Phi} ab(\varphi) = \sup (aB),$$

giving the desired proof.

A case, showing that (7.1) does not, in general, imply (7.3) is illustrated in the next theorem.

THEOREM 9. *A set of subsets of a set, closed under infinite joins and meets, forms a complete lattice, in which the identities* (7.3) *and* (7.3′) *hold.*

Proof. If $x \in \bigcap_{\alpha \in A} \bigcup_{\beta \in B_\alpha} X_{\alpha\beta}$, then $x \in \bigcup_{\beta \in B_\alpha} X_{\alpha\beta}$ for every α and, consequently, $x \in X_{\alpha\varphi(\alpha)}$ for some $\varphi(\alpha) \in B_\alpha$. Then $x \in \bigcap_{\alpha \in A} X_{\alpha\varphi(\alpha)} \subseteq \bigcup_{\varphi \in \Phi} \bigcap_{\alpha \in A} X_{\alpha\varphi(\alpha)}$. Conversely, if, $x \in \bigcup_{\varphi \in \Phi} \bigcap_{\alpha \in A} X_{\alpha\varphi(\alpha)}$, then $x \in \bigcap_{\alpha \in A} X_{\alpha\varphi(\alpha)}$ for some $\varphi \in \Phi$. Thus, for every $\alpha \in A$, there exists an index $\beta \in B_\alpha$, such that $x \in X_{\alpha\beta}$ and, consequently, $x \in \bigcap_{\beta \in B_\alpha} X_{\alpha\beta}$ for all $\alpha \in A$. The latter implies that $x \in \bigcap_{\alpha \in A} \bigcup_{\beta \in B_\alpha} X_{\alpha\beta}$, completing the proof for relation (7.3).

Now, let $x \in \bigcup_{\alpha \in A} \bigcap_{\beta \in B_\alpha} X_{\alpha\beta}$. Then for some $\alpha \in A$ and for all $\beta \in B_\alpha$ we have $x \in X_{\alpha\beta}$. Hence, $x \in X_{\alpha\varphi(\alpha)} \subseteq \bigcup_{\alpha \in A} X_{\alpha\varphi(\alpha)}$ for all $\varphi \in \Phi$, implying that $x \in \bigcap_{\varphi \in \Phi} \bigcup_{\alpha \in A} X_{\alpha\varphi(\alpha)}$. The 'inverse operation' completely establishes the validity of relation (7.3′).

Exercises

1. Prove that the distributivity of the lattice L is equivalent to any of the following properties: (*a*) the inequality $a \leqslant b$ holds, iff $ac \leqslant bc$ and $a + c \leqslant b + c$ for some $c \in L$; (*b*) $(a + b)(c + ab) = ab + bc + ac$; (*c*) $a + bc \geqslant (a + b)c$; (*d*) L is a modular lattice and does not contain the sublattice of Fig. 11.

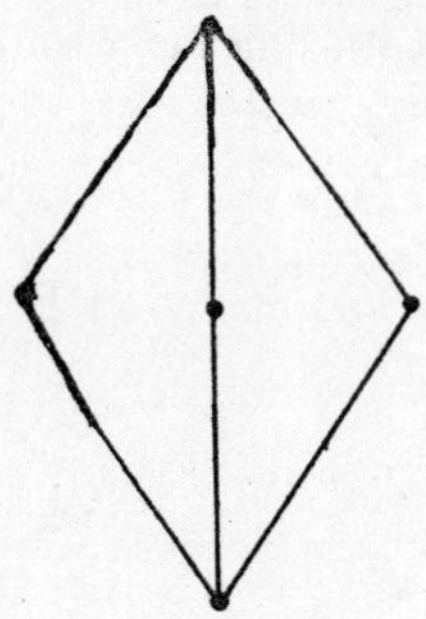

Figure 11.

2. In a distributive lattice the relations $ab \leqslant x \leqslant a + b$ and $x = ax + bx + ab$ are satisfied simultaneously.
3. A modular lattice is distributive, iff the homomorphic image of any of its sublattices, with no non-trivial congruence is either a one- or two-element chain.
4. The ordinal sum of distributive lattices is distributive.
5. If A is the set of atoms of a continuous distributive lattice and $\bar{a} = \sup (A \setminus a)$, then $a\bar{a} = 0$ for all $a \in A$.
6. A distributive lattice of finite length is finite.
7. In a distributive lattice, satisfying the maximum condition, every element has precisely one irreducible $\cap$-representation.
8. If L is a distributive lattice, having a composition series and if $1 = a_1 \dot{+} \cdots \dot{+} a_m = b_1 \dot{+} \cdots \dot{+} b_n$, where a_i and b_j are indecomposable elements, then $m = n$ and when suitably indexed, we have also $a_i = b_i$ for all i.
9. A sublattice of a distributive lattice, generated by a finite number of elements, is finite.
10. A complete distributive lattice, satisfying equality (7.1) is *completely modular*, that is, for any element $\{x_\alpha, y_\alpha \mid \alpha \in \Omega\}$, where $x_\alpha \leqslant y_\beta$ for $\alpha \neq \beta$ there holds $\sup \{x_\alpha\} \cdot \inf \{y_\alpha\} = \sup \{x_\alpha y_\alpha\}$. (Cf. Exercise 5 in Chap. 6.)
11. Let L be a distributive lattice, $a, b \in L$, $a \leqslant b$ and $I = \{x \mid b(a + x) = a\}$. Show that I is an ideal. If the interval $[a, b]$ is simple, then $ad \in I$ implies either $c \in I$ or $d \in I$.
12. A maximal ideal of a distributive lattice is prime. Formulate and prove the dual statement.
13. The ideal a^∇ of a distributive lattice is prime, iff a is a $\cap$-irreducible element.

14. The distributivity of a lattice is equivalent to the distributivity of the lattice of its ideals.

15. Define the free distributive lattice. Show that it exists and is unique.

16. A sublattice of a free distributive lattice, generated by a part of the free generators, is free.

17. For any set $\{L_\alpha \mid \alpha \in \Omega\}$ of distributive lattices, there exists a distributive lattice L, generated by the join $\bigcup_{\alpha \in \Omega} L_\alpha$, and for any collection $\{\varphi_\alpha \mid \alpha \in \Omega\}$ of the homomorphisms of the lattices L_α into any distributive lattice M, there is a homomorphism of L into M, which on each of the L_α coincides with φ_α.

8

BOOLEAN ALGEBRA

A complemented distributive lattice is called a *Boolean algebra* (lattice). By Theorem 6.3 a Boolean algebra is a relatively complemented lattice and by Theorem 7.1 each element a in it has a unique complement a'. Hence, we can say that in Boolean algebra, besides the two binary operations ($\cup$, $\cap$), there is defined a unary operation of complementation.

THEOREM 1. *In every Boolean algebra the following relations hold:*

$$(a + b)' = a'b', \tag{8.1}$$

$$(ab)' = a' + b', \tag{8.2}$$

$$a'' = a, \tag{8.3}$$

$$0' = 1, \tag{8.4}$$

$$1' = 0. \tag{8.5}$$

Proof. The validity of equalities (8.3)–(8.5) is trivial. Equality (8.1) follows from the relations

$$(a + b) + a'b' = (a + a')(a + b') + b = a + b' + b = 1,$$

and

$$(a + b)\, a'b' = aa'b' + a'bb' = 0,$$

obtained by Theorem 7.1, and dually, equality (8.2) holds.

THEOREM 2. *If in a lattice L with zero and unit element, any element a has a unique complement a′, where $(a + b)' = a'b'$ and $(ab)' = a' + b'$, then L is a Boolean algebra.*

Proof. Let us first prove the next preparatory lemma.

LEMMA. *If $x, y \in L$ and $x \leqslant y$, then $x = (x + y')y$ and $y = x + x'y$.* Indeed, taking note of the hypothesis of theorem, we obtain

$$(x + x'y)' + y \geqslant (x + x'y)' + (x + x'y) = 1,$$

and

$$(x + x'y)'\, y = x'\, (x + y')\, y = (x'y)\, (x'y)' = 0.$$

Consequently, $(x + x'y)' = y'$ and, hence, $y = x + x'y$. Furthermore, from the equality $x = xy$, it follows that $x' = x' + y'$, that is, $y' \leqslant x'$. By what has been proved above, we have $x' = y' + yx'$, whence

$$x = (y' + yx')' = (x + y')\, y.$$

Passing to the proof of theorem, we assume that

$$a + c = v = b + c,$$

and

$$ac = u = bc.$$

Taking note of the lemma, we get the equalities

$$\begin{aligned} a + (v' + cu') &= a + (u + cu') + v' = (a + c) + v' = 1 \\ &= (b + c) + v' = b + (u + cu') + v' \\ &= b + (v' + cu'), \end{aligned}$$

and

$$a(v' + cu') = av(v' + cu') = (ac)u' = 0 = (bc)u'$$
$$= bv(v' + cu') = b(v' + cu').$$

Consequently,

$$a = (v' + cu')' = b,$$

and by property (iv) of Theorem 7.1, L is a distributive lattice, giving the desired conclusion.

THEOREM 3. *Let B be a Boolean algebra. For any $a, b \in B$, put*

$$a \boldsymbol{+} b \underset{\text{def}}{=} ab' + a'b,$$

and

$$a \circ b \underset{\text{def}}{=} ab.$$

Then B is an associative, commutative ring with unit element, where $a \boldsymbol{+} a = 0$ and $a \circ a = a$ for all $a \in B$.

Proof. The last statement, associativity of multiplication, presence of unit element and commutativity of both the binary operations, is obvious. Further, by Theorem 1, we have

$$(a \boldsymbol{+} b) \boldsymbol{+} c = (ab' + a'b)c' + (ab' + a'b)'c$$
$$= ab'c' + a'bc' + (a' + b)(a + b')c$$
$$= ab'c' + a'bc' + a'b'c + abc$$
$$= a(b' + c)(b + c') + a'(bc' + b'c)$$
$$= a(bc' + b'c)' + a'(bc' + b'c) = a \boldsymbol{+} (b \boldsymbol{+} c),$$
$$a \boldsymbol{+} 0 = a0' + a'0 = a1 = a,$$

and

$$a \boldsymbol{+} a = aa' + a'a = 0,$$

that is, B forms an additive Abelian group. Finally

$$(a \dotplus b) \circ c = ab'c + a'bc = ac\,(b' + c') + (a' + c')\,bc$$

$$= (ac)\,(bc)' + (ac)'\,(bc) = a \circ c \dotplus b \circ c.$$

THEOREM 4. *Let R be an associative, commutative ring with unit element relative to the operations $\dotplus$ and $\circ$, where $a \circ a = a$ for all $a \in R$. Put*

$$a + b \underset{\text{def}}{=} a \dotplus b \dotplus a \circ b,$$

and

$$ab \underset{\text{def}}{=} a \circ b.$$

Then R is a Boolean algebra B. The ring obtained from an algebra B by Theorem 3 *coincides with R. The application of the described construction to the ring, indicated in Theorem* 3, *leads to the original Boolean algebra B.*

Proof. The commutativity and associativity of the operations $+$ and $\cdot$ are obvious. In addition, we have

$$(a \dotplus a) \dotplus (a \dotplus a) = (a \dotplus a) \circ (a \dotplus a) = a \dotplus a,$$

whence,

$$a \dotplus a = 0.$$

Taking note of this equality, we get

$$(a + b)\,a = (a \dotplus b \dotplus a \circ b) \circ a = a \circ a \dotplus a \circ b \dotplus a \circ b = a,$$

and

$$a + ab = a \dotplus a \circ b \dotplus a \circ (a \circ b) = a.$$

By Theorem 4.2, B forms a lattice. Clearly, $a \cdot 1 = a$ and $a0 = 0$ for all $a \in B$, that is, the lattice B has zero and unit element. From the equalities

$$a\,(1 \dotplus a) = a \circ 1 \dotplus a \circ a = 0,$$

and

$$a + (1 \dotplus a) = a \dotplus (1 \dotplus a) \dotplus a \circ (1 \dotplus a) = 1,$$

it follows that B is a complemented lattice. Its distributivity is verified in the following way:

$$(a + b)\, c = (a \dotplus b \dotplus ab) \circ c = ac \dotplus bc \dotplus ac \circ bc = ac + bc.$$

The concluding assertion of the Theorem follows from the equalities

$$\begin{aligned} ab' + a'b &= a \circ (1 \dotplus b) \dotplus (1 \dotplus a) \circ b \dotplus a \circ (1 \dotplus b) \circ b \circ (1 \dotplus a) \\ &= a \dotplus ab \dotplus b \dotplus ab = a \dotplus b, \end{aligned}$$

and

$$\begin{aligned} &(ab' + a'b)\,(ab)' + (ab' + a'b)'\,ab \\ &\quad = (ab' + a'b)\,(a' + b') + (a' + b)\,(a + b')\,ab \\ &\quad = ab' + a'b + ab = a\,(b' + b) + (a' + a)\,b = a + b. \end{aligned}$$

THEOREM 5. (Glivenko–Stone). *The completion by cuts of a Boolean algebra is a complete Boolean algebra.*

Proof. Given a Boolean algebra M and let L be its completion by cuts as described in Theorem 3.7. It is understood that the elements of L serve as $\Delta\nabla$–closed elements of the lattice P of all subsets of M. However, let us now agree that $\phi \,\overline{\in}\, L$. If $A \in P$, then put

$$A' = \{a' \mid a \in A\}.$$

LEMMA 1. $A'^{\nabla\Delta} = (A^{\Delta\nabla})'$.

Indeed, $x \in A'^{\nabla\Delta}$ means the same as $x \geqslant y$ for all $y \in A'^{\nabla}$. However, the latter is equivalent to the inequality $x' \leqslant y'$ for all $y' \in A^{\Delta}$. Hence, $x \in A'^{\Delta\nabla}$ is equivalent to $x' \in A^{\Delta\nabla}$ or, what is the same, $x \in (A^{\Delta\nabla})'$.

LEMMA 2. *If* $x \leqslant y$ *and* $y \in A^{\Delta\nabla}$, *then* $x \in A^{\Delta\nabla}$.

For proof, it is sufficient to note that $x \leqslant y \leqslant z$ for all $z \in A^{\Delta}$.

LEMMA 3. *If* $A \in L$, *then* $A^{\Delta} \cap A' = 1$.

In fact, if $x \in A^{\Delta} \cap A'$, then $x' \in A$, whence, $x' = xx' = 0$ and, consequently, $x = 1$. On the other hand, by Lemma 2, $0 \in A$ and, consequently, $1 = 0' \in A^{\Delta} \cap A'$.

LEMMA 4. *If $A \in L$, then $A \cap A'^{\nabla} = 0$.*

Indeed, by Lemma 2, $0 \in A \cap A'^{\nabla}$. However, if $x \in A \cap A'^{\nabla}$, then $x = xx' = 0$.

LEMMA 5. *If $A \in L$, then A'^{∇} serves as a unique complement of the element A in the lattice L.*

Indeed, by property (iv) of Theorem 1.6, $A'^{\nabla} \in L$. Furthermore, making use of Theorem 3.5, property (v) of Theorem 1.6 and Lemmas 1, 3 and 4, we get

$$\sup_L \{A, A'^{\nabla}\} = (A \cup A'^{\nabla})^{\Delta\nabla} = (A^{\Delta} \cap A'^{\nabla\Delta})^{\nabla} = (A^{\Delta} \cap (A^{\Delta\nabla})')^{\nabla}$$
$$= (A^{\Delta} \cap A')^{\nabla} = 1^{\nabla} = M,$$

and

$$\inf_L \{A, A'^{\nabla}\} = A \cap A'^{\nabla} = 0.$$

Now, let $A, B \in L$, $\sup_L \{A, B\} = M$ and $\inf_L \{A, B\} = 0$. By Theorem 3.5, we have

$$(A \cup B)^{\Delta\nabla} = M,$$

and

$$A \cap B = 0.$$

Properties (iii) and (v) of Theorem 1.6 yield

$$1 = M^{\Delta} = (A \cup B)^{\Delta\nabla\Delta} = A^{\Delta} \cap B^{\Delta}.$$

If $b \in B$, then from Lemma 2 if follows that for all $a \in A$

$$ab \in A \cap B = 0.$$

Hence,

$$b = (a + a')\, b = a'b,$$

that is, $b \leqslant a'$ for all $a \in A$. Thus, $B \subseteq A'^{\nabla}$. However, if $x \in A'^{\nabla}$, then $x \leqslant a'$ for all $a \in A$. Hence, $x' \in A^{\Delta}$. If $y \in B^{\Delta}$, then

$$x' + y \in A^{\Delta} \cap B^{\Delta},$$

whence

$$x' + y = 1.$$

However, then, $x = x(x' + y) = xy$, that is, $x \leqslant y$ for all $y \in B^{\vartriangle}$. Consequently, $x \in B^{\vartriangle\triangledown} = B$, that is, $A'^{\triangledown} \subseteq B$. Hence, $A'^{\triangledown} = B$, and the uniqueness of the complement A' is proved.

The validity of Theorm 5 is immediate from Lemma 5 and Theorem 2, by taking note of the relations

$$\begin{aligned}\sup_L \{A'^{\triangledown}, B'^{\triangledown}\} &= (A'^{\triangledown} \cup B'^{\triangledown})^{\vartriangle\triangledown} \\ &= (A'^{\triangledown\vartriangle} \cap B'^{\triangledown\vartriangle})^{\triangledown} = (A' \cap B')^{\triangledown} \\ &= (A \cap B)'^{\triangledown} = (\inf_L \{A, B\})'^{\triangledown},\end{aligned}$$

and

$$\begin{aligned}\inf_L \{A'^{\triangledown}, B'^{\triangledown}\} &= A'^{\triangledown} \cap B'^{\triangledown} = (A' \cup B')^{\triangledown} \\ &= (A' \cup B')^{\triangledown\vartriangle\triangledown} = (A \cup B)'^{\triangledown\vartriangle\triangledown} \\ &= ((A \cup B)^{\vartriangle\triangledown})'^{\triangledown} = (\sup_L \{A, B\})'^{\triangledown},\end{aligned}$$

obtained by means of Theorem 3.5, Theorem 1.6 and Lemma 1.

It remains to verify that the presence of a complement is essential for Theorem 5 to be valid. Let $C = \{u, v\}$ be a two-element chain, and let $A = \{a_1, a_2, \ldots\}$ and $B = \{b_1, b_2, \ldots\}$ be two copies of a natural number, A^* and B^* be the duals of the chain, $\tilde{A}$ be the ordinal sum of A and A^* and let $\tilde{B}$ be the ordinal sum of B and B^* (Figure 12). In the lattice $\tilde{A} \times \tilde{B} \times C$ we consider the subset

$$M = (\tilde{A}, B^*, v) \cup (A, \tilde{B}, u).$$

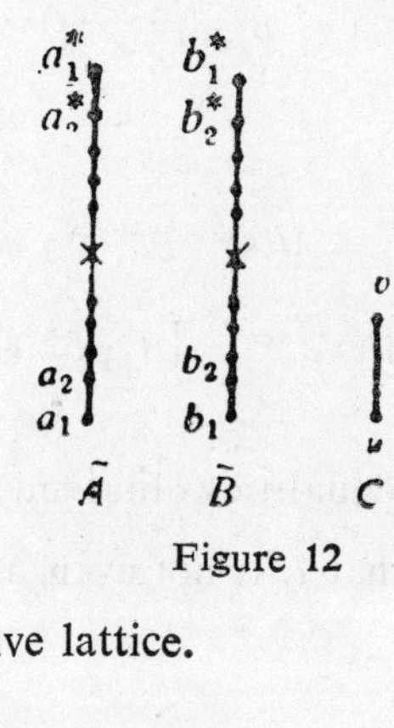

Figure 12

Clearly, M is a distributive lattice.

As above, we denote by L a completion by cuts of M, that is, a lattice of $\Delta\nabla$-closed subsets of the set M. Further, consider the sets

$$D = (A, B, u),$$

$$S = (a_1, b_1^*, v)^{\Delta\nabla} = (\tilde{A}, b_1^*, v)^{\nabla}$$
$$= (a_1, \tilde{B}, C) \cap M = (a_1, B^*, v) \cup (a_1, \tilde{B}, u),$$

$$T = (a_1, b_1^*, u)^{\Delta\nabla} = [(\tilde{A}, b_1^*, C) \cap M]^{\nabla}$$
$$= [(\tilde{A}, b_1^*, v) \cup (A, b_1^*, u)]^{\nabla} = (a_1, \tilde{B}, u),$$

and

$$U = (a_1, B, u).$$

Clearly, $S, T \in L$. Since

$$D^{\Delta\nabla} = ((A^*, B^*, C) \cap M)^{\nabla} = (A^*, B^*, v)^{\nabla}$$
$$= (A, B, C) \cap M = (A, B, u) = D,$$

hence, $D \in L$. Moreover, we have

$$D \cap S = D \cap (a_1, \tilde{B}, C) = U = D \cap T,$$

and by Theorem 1.6 we get

$$(D \cup S)^{\Delta\nabla} = (D^{\Delta} \cap S^{\Delta})^{\nabla}$$
$$= [(A^*, B^*, C) \cap (\tilde{A}, b_1^*, v) \cap M]^{\nabla} = (A^*, b_1^*, v)^{\nabla},$$

and

$$(D \cup T)^{\Delta\nabla} = (D^{\Delta} \cap T^{\Delta})^{\nabla} = [(A^*, B^*, C) \cap (\tilde{A}, b_1^*, C) \cap M]^{\nabla}$$
$$= [(A^*, b_1^*, C) \cap M]^{\nabla} = [A^*, b_1^*, v)^{\nabla}.$$

Since $S \underset{\neq}{\supset} T$, therefore the equalities obtained together with Theorem 3.5 and Theorem 6.1 show that L is not even a modular lattice†.

†This example is due to Funayama.

Finally, we shall examine in a complete Boolean algebra the position of infinite distributive laws (7.1)–(7.3), considered in Chapter 7.

THEOREM 6. *In every complete Boolean algebra, the following equalities hold*:

$$a \cdot \sup B = \sup (aB) \quad \text{and} \quad a + \inf B = \inf (a + B).$$

Proof. It is clear that $a \cdot \sup B \geqslant ab$ for all $b \in B$. If $v \geqslant ab$ for all $b \in B$, then

$$a' + v \geqslant a' + ab = (a' + a)(a' + b) = a' + b \geqslant b$$

for all $b \in B$. Hence, $a' + v \geqslant \sup B$ and, consequently,

$$a \cdot \sup B \leqslant a(a' + v) = av \leqslant v.$$

Thus,

$$a \cdot \sup B = \sup (aB).$$

The validity of second equality follows by duality.

The next conclusion is immediate from Theorems 5 and 6 of the present chapter and Theorem 3.7.

COROLLARY. *In any Boolean algebra, the equalities indicated in Theorem 6 hold whenever universal bounds entering it exist.*

THEOREM 7. *The following properties of a complete Boolean algebra B are equivalent*:

(i) *B is isomorphic with the lattice of all subsets of some set M*;

(ii) *the identity* (7.3) *holds in B*;

(iii) *the identity* (7.3′) *holds in B*;

(iv) *B is atomic.*

Proof. From Theorem 7.9 follow the implications (i) ⇒ (ii) and (i) ⇒ (iii).

(ii) ⇒ (iv). Denote by Φ the set of all mappings of B into a two-element set $\{0, 1\}$. For any $b \in B$ and $\varphi \in \Phi$ put $b^0 = b$, $b^1 = b'$,

$$\theta_\varphi = \inf_{b \in B} \{b^{\varphi(b)}\} \quad \text{and} \quad \Phi' = \{\varphi \mid \varphi \in \Phi, \theta_\varphi \neq 0\}.$$

If $0 < x \leqslant \theta_\varphi$, then $x \leqslant x^{\varphi(x)}$. If $\varphi(x) = 1$, then $x \leqslant xx' = 0$, violating the hypothesis. Consequently, $\varphi(x) = 0$. Hence, $x \leqslant \theta_\varphi \leqslant x^0 = x$, that is, $x = \theta_\varphi$. Thus, θ_φ is an atom for every $\varphi \in \Phi'$. Using (7.3), we get

$$1 = \inf_B (b + b') = \inf_B \sup_{\{0,1\}} \{b^i\}$$

$$= \sup_{\varphi\in\Phi} \inf_B \{b^{\varphi(b)}\} = \sup_{\varphi\in\Phi} \theta_\varphi = \sup_{\varphi\in\Phi'} \theta_\varphi.$$

Further, if, $0 \neq b \in B$, then by Theorem 6 we have

$$b = b \sup_{\varphi\in\Phi'} \theta_\varphi = \sup_{\varphi\in\Phi'} b\theta_\varphi.$$

For some $\varphi \in \Phi'$ we have $0 < b\theta_\varphi \leqslant \theta_\varphi$. Since θ_φ is an atom, it follows that $b\theta_\varphi = \theta_\varphi$. Consequently, $\theta_\varphi \leqslant b$ as we wished to show.

(iv) ⇒ (i). Let M be the set of all atoms of B and let P be the lattice of all subsets of M. If $b \in B$, we put

$$\varphi(b) = \begin{cases} \{p \mid p \in M,\ p \leqslant b\}, & \text{if} \quad b \neq 0, \\ \phi & \text{if} \quad b = 0. \end{cases}$$

Then φ maps B into P. We define the mapping of P into B, by putting

$$\psi(X) = \begin{cases} \sup X, & \text{if } X \text{ is not empty}, \\ 0, & \text{if } X = \phi. \end{cases}$$

By Theorem 6.20, $\psi\varphi(b) = b$ for all $b \in B$. Consequently, ψ maps P onto B. Assume that $\psi(X) = \psi(Y)$ and $y \in Y$. By Theorem 6, we have

$$y = y\psi(Y) = y\psi(X) = y \sup X = \sup(yX),$$

whence

$$0 < xy \leqslant x, y$$

for some $x \in X$. Since x and y are atoms, it follows that

$$y = xy = x \in X.$$

Hence, $Y \subseteq X$ and, similarly $X \subseteq Y$. Consequently, ψ is a one-to-one correspondence. Clearly, ψ is isotone.

(iii) $\Rightarrow$ (i). It is sufficient to note that this implication is just the dual of the implication (ii) $\Rightarrow$ (i) already proved.

From Theorem 7 the next conclusion is immediate.

COROLLARY. *A finite Boolean algebra is isomorphic with a lattice of all subsets of some finite set.*

Finally, it remains to be verified that the equality (7.3) does not hold in every complete Boolean algebra. To this end, consider the set P of all open subsets in a segment [0, 1] of real numbers, regarded as a topological space with the usual topology. Since the void set and any set union of open sets are open, it follows dually from Theorem 3.1, that P is a complete lattice. If $a \in P$, we put $Ca = [0, 1]\backslash a$, and denote by $\bar{a}$ the closure of a. In addition, we put $a' = C\bar{a}$ and $\varphi(a) = a''$. Clearly, $a' \in P$.

LEMMA 1. $(a \cup b)' = a' \cap b'$.

Indeed, from the well known closure properties† (see, for example, P. S. Alexandrov, *Introduction to the General Theory of Sets and Functions*, Gostekhizdat, 1948, p. 237, Theorem 6) it follows that

$$(a \cup b)' = C\,(\overline{a \cup b}) = C\,(\bar{a} \cup \bar{b}) = C\bar{a} \cap C\bar{b} = a' \cap b'.$$

LEMMA 2. *If* $a \subseteq b$, then $a' \supseteq b'$.

In fact, if $a \subseteq b$, then $a \cup b = b$. Applying Lemma 1, we get

$$b' = (a \cup b)' = a' \cap b',$$

whence, $a' \supseteq b'$.

LEMMA 3. $a' \subseteq a'''$.

Indeed, from the obvious inclusion $C\bar{\bar{a}} \subseteq \overline{C\bar{a}}$ it follows that $C\overline{C\bar{a}} \subseteq CC\bar{a} = \bar{a}$. Therefore,

$$a'' = \overline{Ca'} = C\overline{C\bar{a}} \subseteq \bar{a},$$

and, consequently, $\bar{a}'' \subseteq \bar{a}$. Hence

$$a' = C\bar{a} \subseteq C\bar{a}'' = a'''.$$

†J. L. Kelley, *General Topology*, Van Nostrand, 1955.

LEMMA 4. $a \subseteq a''$.

In fact, since a is open, we have

$$Ca = \overline{Ca} \subseteq \overline{C\bar{a}},$$

whence,

$$a = CCa \subseteq C\overline{C\bar{a}} = \overline{Ca'} = a''.$$

LEMMA 5. *φ is a closure operator on the lattice P.*

In fact, the relation $a \subseteq \varphi(a)$ follows from Lemma 4, and the implication $a \subseteq b \Rightarrow \varphi(a) \subseteq \varphi(b)$ from Lemma 2. Finally, from Lemmas 4, 3 and 2 it follows that

$$\varphi(a) = a'' \subseteq (a'')'' = \varphi(\varphi(a)) = (a''')' \subseteq a'' = \varphi(a).$$

By Lemma 5, from Theorem 3.5 it follows that the φ-closed elements of the lattice P form a complete lattice B. Lemmas 3 and 4 imply that $a' = a''' \in B$. By Lemma 1 and Theorem 3.5, for all $a \in B$ the relation

$$\overline{a \cup a'} = \overline{a \cup C\bar{a}} = \bar{a} \cup \overline{C\bar{a}} = [0, 1]$$

implies the equalities

$$a + a' = (a \cup a')'' = (C\overline{a \cup a'})' = (\phi)' = C\phi = [0, 1],$$

and

$$a \cdot a' = a \cap a' \subseteq a'' \cap a' = (a' \cup a)' = C\overline{a' \cup a} = \phi.$$

If $a + c = [0, 1]$ and $ac = \phi$ for some $c \in B$, then the second of these relations yields $a \cap c = \phi$, whence, $a \subseteq Cc$ and since c is open, $\bar{a} \subseteq Cc$. Consequently, $c = CCc \subseteq C\bar{a} = a'$. From the first relation by Lemma 1, we get

$$(a' \cap c')' = (a \cup c)'' = [0, 1].$$

Since $a' \cap c' \in B$, we have

$$a' \cap c' = (a' \cap c')'' = [0, 1]' = \phi.$$

Therefore,

$$C\bar{a} = a' \subseteq Cc' = CC\bar{c}.$$

However, then

$$\bar{a} \supseteq C\bar{c} = \bar{c}',$$

whence,

$$\bar{a} \supseteq c',$$

and, consequently,

$$a' = C\bar{a} \subseteq C\bar{c'} = c'' = c.$$

Thus, every element a of the lattice B has a unique complement a'. Since Theorem 3.5 and Lemma 1 imply the equalities

$$(a + b)' = (a \cup b)''' = (a' \cap b')'' = a' \cap b' = a'b',$$

and

$$(ab)' = (a'' \cap b'')' = (a' \cup b')'' = a' + b',$$

it follows by Theorem 2, that B forms a Boolean algebra. It remains to note that any non-trivial element in B contains an open interval. But, every such interval is a φ-closed set. Consequently, B contains no atoms and, by Theorem 7, the relation (7.3) is not satisfied there.

Exercises

1. For any elements a and b in a Boolean algebra, the following assertions are equivalent:

   (a) $a \leqslant b$; (b) $ab' = 0$; (c) $a' + b = 1$.

2. A collection of complemented elements of a distributive lattice forms a sublattice and is a Boolean algebra.
3. A collection of elements of a modular lattice, having a unique complement, forms a sublattice and is a Boolean algebra.
4. Every distributive lattice is a sublattice of some Boolean algebra.
5. Any homomorphism φ of a Boolean algebra B onto a Boolean algebra B' is *Boolean* (that is, $\varphi(b') = (\varphi(b))'$ for all $b \in B$).

6. Generalize Theorems 3 and 4 for a distributive relatively complemented lattice with zero.

7. Every congruence relation on a Boolean algebra is a congruence on the corresponding ring and conversely (cf. Theorem 3).

8. Show that the correspondence between the ideals of a Boolean algebra and the ideals of the corresponding ring is one-to-one.

9. All ideals of a Boolean algebra are principal, iff the algebra is finite.

10. Any ideal of a Boolean algebra coincides with the intersection of all prime ideals containing it.

11. A distributive lattice is a Boolean algebra, iff each of its prime ideals is maximal.

12. The ideal I of a Boolean algebra is maximal, iff $a \in I$ and $a' \overline{\in} I$ are equivalent.

13. A Boolean algebra, having a composition series, is finite.

14. If L is a complemented lattice and if $ab = 0$ implies $b \leqslant c$ for each complement c of the element a, then L is a Boolean algebra.

15. If every element of a complete atomic lattice L has precisely one complement, then L is a Boolean algebra.

16. Define the free Boolean algebra. Show that it exists and is unique.

17. The least subalgebra of a free Boolean algebra, containing a part of the free generators, is a free Boolean algebra.

18. For any set $\{B_\alpha \mid \alpha \in \Omega\}$ of a Boolean algebra, there exists a Boolean algebra B, generated (as a Boolean algebra) by the join $\bigcap_{\alpha \in \Omega} B_\alpha$ and for any collection of Boolean homomorphisms $\{\varphi_\alpha \mid \alpha \in \Omega\}$ of the algebras B_α into any Boolean algebra M there exists a Boolean homomorphism of B into M, coinciding with φ_α on each B_α.

BIBLIOGRAPHY

BIRKHOFF, G., *Lattice Theory*, American Mathematical Society, Providence, 3rd edition, 1967.

CARVAILLO, M., *Monographie des treillis et algèbre de Boole*, Gauthier-Villars, Paris, 1966.

DONNELLAN, T., *Lattice Theory*, Pergamon Press, 1968.

DUBREIL-JACOTIN, M. L., LESIEUR, L. and CROISOT, R., *Lecons sur la théorie des treillis, des structures algébriques ordonnées et des treillis géométriques*, Gauthier-Villars, Paris, 1953.

GERICKE, H., *Theorie der Verbände*, Mannheim, 1967.

GLUKHOV, M. M., STELLETSKII, I. V. and FOFANOVA, T. S., *Lattices, Results in Science, Algebra, Geometry and Topology*, Moscow, 1970, pp. 101–55.

GOODSTEIN, H., *Boolean Algebra*, Pergamon Press, 1963.

HALMOS, P., *Lectures on Boolean Algebra*, Van Nostrand, 1963.

HERMES, H., *Einführung in die Verbandstheorie*, Springer-Verlag, 1967.

MAEDA, F., *Kontinuierliche Geometrien*, Springer-Verlag, 1958.

MORGADO, J., *Introducão a teoria des reculados*, Recife, 1962.

NACHBIN, L., *Topology and Order*, Van Nostrand, 1965.

NEUMANN, J. VON, *Continuous Geometry*, Princeton, 1960.

RUTHERFORD, D., *Introduction to Lattice Theory*, Oliver and Boyd, 1965.

SIKORSKII, R., *Boolean Algebra*, Springer-Verlag, 2nd edition, 1964.

SKORNYAKOV, L. A., *Complemented Modular Lattice and Regular Rings*, Fizmatgiz, 1961. (Engl. Trans. Oliver and Boyd, 1964).

SKORNYAKOV, L. A., *Lattice Theory, Results in Science and Algebra 1964*, Moscow, 1966, pp. 237–278.

SZÁSZ, G., *Introduction to Lattice Theory*, Akadémiai Kiadó, Budapest 1959; (Engl. Trans. Academic Press, 1963).

VLADIMIROV, D. A., *Boolean Algebra* (Russian), Nauka, 1969.

INDEX

A

Algebra, Boolean 129
Atom 109
Axiom of choice 14
Axis of perspectivity 96

B

Birkhoff theorem 123
Bound
  greatest lower 5
  least upper 5

C

Cantor-Bernstein theorem 23
Cardinal sum 8
Chain 8
  bounded 99
  maximal 14
Cofinal subchain 19
Complement 47
Completion by cuts 36
Composition series 8
Condition
  descending chain 12
  induction 12
  minimum 12
Cone
  lower 6
  upper 6
Congruence 45
  permutable 49
Crawley-Dean theorem 89

D

Davis theorem 54
Diagonal 1
Dilworth theorem 49
Dilworth-Dean theorem 69
Dual expression 4
  of a poset 4
Duality principle theorem 5

E

Element
  distinguished 15
  greatest 4
  indecomposable 103
  least 4
  maximal 4
  minimal 4
    φ-closed 32
    irreducible 108

Elements
  comparable 4
  direct sum of 93
  independent 93
  perspective 96
Equivalence 2
Extension
  complete free 58
  free 58
  $(\mathfrak{A}, \mathfrak{P})$-free 58
  of given expansion 11

F

Filter 50
  complete 55
  kernel 50
  prime 56
  principal 50
  standard 54
Function, dimension 98

G

Generators, free 72
Glivenko-Stone theorem 133
Grätzer-Schmidt theorem 51, 53

H

Hashimoto theorem 47, 119
Hausdorff's theorem 15
Homomorphism 44
  Boolean 141
  complete 55
  kernel of 46
  $\cup$-homomorphism 56
  φ-closure 32
  natural 46
  theorem 46

I

Ideal 49
  complete 55
  kernel 50
  prime 56
  principal 50
  standard 50
Identity 1
Infinite distributive law 124
Initial segment 14
Interval 7
  simple 7
Irredundant $\cap$-representation 108
Isomorphic posets 4
Isomorphism 45

K

Kogalovski theorem 30
Kolibiar-Hashimoto-Grätzer-Schmidt
  theorem 120
Kuratowski-Zorn lemma 15

L

Lattice 41
  atomic 109
  complemented 49
  complete 29
  completely modular 127
  continuous 111
  Dedekind 90
  distributive 117
  free 59
    complete 86
    modular 112
  -morphism 44
  modular 90
  quotient 40
  relatively complemented 47
Lemma, Kuratowski-Zorn 15

M

Mapping
  isotone 4
  natural 2
  perspective 97

O

Operator, closure 32
Order 3
Ordinals 13
  limit 14
Ore theorem 103

P

Partition 2
Principal, duality 5
Product
  direct 8, 20
  free 59
  lexicographic 21
  ordered 21

R

Relation 1
  antisymmetric 1
  identity 1
  induced 1
  reflexive 1
  residue class of 2
  symmetric 1
  transitive 1
Representation
  $\cup$-canonical 59
  $\cap$-canonical 59

S

Set
  direct 111
  equivalent 24
  factor 2
  partially ordered 3
    cardinally indecomposable 9
    ordinally indecomposable 9
    trivial 4
  power of 24
  well ordered 13
Stone theorem 122
Sublattice 41
  complete 55
  completely generated by a set 55
  convex 41
  generated by a set 41
Sum
  ordered 8
  ordinal 8
Subset
  having Zorn's property 15
  multiplicative 58
  summable 58
Subword 59

T

Theorem
  Birkhoff 123
  Cantor-Bernstein 23
  Comparison of sets 21, 24
  Crawley-Dean 89
  Davis 54
  Dilworth 49
  Dilworth-Dean 69
  duality principle 5
  fixed point 29
  Glivenko-Stone 133
  Grätzer-Schmidt 51, 53
  Hashimoto 47, 119
  Kolibiar-Hashimoto-
    Grätzer-Schmidt 120
  homomorphism 46
  Ore 103
  Stone 122
  Whitman 72

U

Unit 4

W

Whitman theorem 72
Word 59
  $\cup$-irreducible 59
  $\cap$-irreducible 59
  reduced 80
    completely 80
  weight of 59

Z

Zermelo's theorem 15
Zero 4

LIST OF SPECIAL SYMBOLS

$\mathfrak{P}(\mathfrak{M})$ 3

$\sup_P A$, $\inf_P A$ 5

$A^{\vartriangle}$ $A^{\triangledown}$ 6

$FL(n)$ 72

$FML(n)$ 112

$P_{a \to b;\, c}$ 97

$\dot{+}$ 93

$\leqslant_P$ 8